KARMANN GHIA
1955-1982

Compiled by
R.M. Clarke

ISBN 0 948207 43 4

Booklands Books Ltd.
PO Box 146, Cobham, KT11 1LG
Surrey, England

Printed in Hong Kong

BROOKLANDS BOOKS

BROOKLANDS ROAD TEST SERIES

AC Ace & Aceca 1953-1983
Alfa Romeo Alfasud 1972-1984
Alfa Romeo Alfetta Coupes GT. GTV. GTV6 1974-1987
Alfa Romeo Giulia Berlinas 1962-1976
Alfa Romeo Giulia Coupes Gold Portfolio 1963-1976
Alfa Romeo Giulia Coupes 1963-1976
Alfa Romeo Giulietta Gold Portfolio 1954-1965
Alfa Romeo Spider Gold Portfolio 1966-1991
Alfa Romeo Spider 1966-1990
Allard Gold Portfolio 1937-1959
Alvis Gold Portfolio 1919-1967
American Motors Muscle Cars 1966-1970
Armstrong Siddeley Gold Portfolio 1945-1960
Aston Martin Gold Portfolio 1972-1985
Austin Seven 1922-1982
Austin A30 & A35 1951-1962
Austin Healey 100 & 100/6 Gold Portfolio 1952-1959
Austin Healey 3000 Gold Portfolio 1959-1967
Austin Healey Sprite 1958-1971
Avanti 1962-1990
BMW Six Cylinder Coupes 1969-1975
BMW 1600 Col. 1 1966-1981
BMW 2002 1968-1976
BMW 316, 318, 320 Gold Portfolio 1975-1990
BMW 320, 323, 325 Gold Portfolio 1977-1990
Buick Automobiles 1947-1960
Buick Muscle Cars 1965-1970
Buick Riviera 1963-1978
Cadillac Automobiles 1949-1959
Cadillac Automobiles 1960-1969
Cadillac Eldorado 1967-1978
High Performance Capris Gold Portfolio 1969-1987
Chevrolet Camaro SS & Z28 1966-1973
Chevrolet Camaro & Z-28 1973-1981
High Performance Camaros 1982-1988
Camaro Muscle Portfolio 1967-1973
Chevrolet 1955-1957
Chevrolet Corvair 1959-1969
Chevrolet Impala & SS 1958-1971
Chevrolet Muscle Cars 1966-1971
Chevelle and SS 1964-1972
Chevy Blazer 1969-1981
Chevy EL Camino & SS 1959-1987
Chevy II Nova & SS 1962-1973
Chrysler 300 Gold Portfolio 1955-1970
Citroen Traction Avant Gold Portfolio 1934-1957
Citroen DS & ID 1955-1975
Citroen SM 1970-1975
Citroen 2CV 1949-1988
Shelby Cobra Gold Portfolio 1962-1969
Cobras and Cobra Replicas Gold Portfolio 1962-1989
Cobras & Replicas 1962-1983
Chevrolet Corvette Gold Portfolio 1953 1962
Corvette Stingray Gold Portfolio 1963-1967
Chevrolet Corvettee Gold Portfolio 1968-1977
High Performance Corvettes 1983-1989
Daimler SP250 Sport & V-8250 Saloon Gold Portfolio 1959-1969
Datsun 240Z 1970-1973
Datsun 280Z & ZX 1975-1983
De Tomaso Collection No.1 1962-1981
Dodge Charger 1966-1974
Dodge Muscle Cars 1967-1970
Excalibur Collection No.1 1952-1981
Facel Vega 1954-1964
Ferrari Cars 1946-1956
Ferrari Dino 1965-1974
Ferrari Dino 308 1974-1979
Ferrari 308 & Mondial 1980-1984
Ferrari Collection No.1 1960-1970
Fiat-Bertone X1/9 1973-1988
Fiat Pininfarina 124 + 2000 Spider 1968-1985
Ford Automobiles 1949-1959
Ford Bronco 1966-1977
Ford Bronco 1978-1988
Ford Consul. Zephyr Zodiac MkI & II 1950-1962
Ford Cortina 1600E & GT 1967-1970
Ford Fairlane 1955-1970
Ford Falcon 1960-1970
Ford GT40 Gold Portfolio 1964-1987
Ford RS Escorts 1968-1980
Ford Zephyr Zodiac Executive MkIII & MkIV 1962-1971
High Performance Capris Gold Portfolio 1969-1987
High Performance Escorts Mk1 1968-1974
High Performance Escorts Mk II 1975-1980
High Performance Escorts 1980-1985
High Performance Escorts 1985-1990
High Performance Fiestas 1979-1991
High Performance Mustangs 1982-1988
Holden 1948-1962
Honda CRX 1983-1987
Hudson & Railton 1936-1940
Jaguar and SS Gold Portfolio 1931-1951
Jaguar XK120 XK140 XK150 Gold Portfolio 1948-1960
Jaguar MkVII VIII IX X 420 Gold Portfolio 1950-1970
Jaguar Cars 1961-1964
Jaguar Mk2 1959-1969
Jaguar E-Type Gold Portfolio 1961-1971
Jaguar E-Type 1966-1971
Jaguar E-Type V-12 1971-1975
Jaguar XJ12 XJ5.3 V12 Glold Portfolio 1972-1990
Jaguar XJ6 Series II 1973-1979
Jaguar XJ6 Series III 1979-1986
Jaguar XJS Gold Portfolio 1975-1990
Jeep CJ5 & CJ6 1960-1976
Jeep CJ5 & CJ7 1976-1986
Jensen Cars 1946-1967
Jensen Cars 1967-1979
Jensen Interceptor Gold Portfolio 1966-1986
Jensen Healey 1972-1976
Lamborghini Cars 1964-1970
Lamborghini Countach Col No.1 1971-1982
Lamborghini Countach & Urraco 1974-1980
Lamborghini Countach & Jalpa 1980-1985
Lancia Fulvia Gold Portfolio 1963-1976
Lancia Stratos 1972-1985
Land Rover Series I 1948-1958
Land Rover Series II & IIa 1958-1971
Land Rover Series III 1971-1985
Land Rover 90 & 110 1983-1989
Lincoln Gold Portfolio 1949-1960
Lincoln Continental 1961-1969
Lincoln Continental 1969-1976
Lotus and Caterham Seven Gold Portfolio 1957-1989
Lotus Cortina Gold Portfolio 1963-1970
Lotus Elan Gold Portfolio 1962-1974
Lotus Elan Collection No.2 1963-1972
Lotus Elite 1957-1964
Lotus Elite & Eclat 1974-1982
Lotus Turbo Esprit 1980-1986
Lotus Europa Gold Portfolio 1966-1975
Marcos Cars 1960-1988
Maserati 1965-1970
Maserati 1970-1975
Mazda RX-7 Collection No.1 1978-1981
Mercedes 190 & 300SL 1954-1963
Mercedes 230/250/280SL 1963-1971
Mercedes Benz SLs & SLCs Gold Portfolio 1971-1989
Mercedes Benz Cars 1949-1954
Mercedes Benz Cars 1954-1957
Mercedes Benz Cars 1957-1961
Mercedes Benz Competion Cars 1950-1957
Mercury Muscle Cars 1966-1971
Metropolitan 1954-1962
MG TC 1945-1949
MG TD 1949-1953
MG TF 1953-1955
MG Cars 1959-1962
MGA & Twin Cam Gold Portfolio 1955-1962
MGB MGC & V8 Gold Portfolio 1962-1980
MGB Roadsters 1962-1980
MGB GT 1965-1980
MG Midget 1961-1980
Mini Cooper Gold Portfolio 1961-1971
Mini Moke 1964-1989
Mini Muscle Cars 1961-1979
Mopar Muscle Cars 1964-1967
Morgan Three-Wheeler Gold Portfolio 1910-1952
Morgan Cars 1960-1970
Morgan Cars Gold Portfolio 1968-1989
Morris Minor Collection No.1
Mustang Muscle Cars 1967-1971
Oldsmobile Automobiles 1955-1963
Old's Cutlass & 4-4-2 1964-1972
Oldsmobile Muscle Cars 1964-1971
Oldsmobile Toronado 1966-1978
Opel GT 1968-1973
Packard Gold Portfolio 1946-1958
Pantera Gold Portfolio 1970-1989
Panther Gold Portfolio 1972-1990
Plymouth Barracuda 1964-1974
Plymouth Muscle Cars 1966-1971
Pontiac Tempest & GTO 1961-1965
Pontiac Firebird and Trans-Am 1973-1981
High Performance Firebirds 1982-1988
Pontiac Fiero 1984-1988
Pontiac Muscle Cars 1966-1972
Porsche 356 1952-1965
Porsche Cars in the 60's
Porsche Cars 1960-1964
Porsche Cars 1964-1968
Porsche Cars 1968-1972
Porsche Cars 1972-1975
Porsche Turbo Collection No.1 1975-1980
Porsche 911 1965-1969
Porsche 911 1970-1972
Porsche 911 1973-1977
Porsche 911 Carrera 1973-1977
Porsche 911 Turbo 1975-1984
Porsche 911 SC 1978-1983
Porsche 914 Gold Portfolio 1969-1976
Porsche 914 Collection No.1 1969-1983
Porsche 924 Gold Portfolio 1975-1988
Porsche 928 1977-1989
Porsche 944 1981-1985
Range Rover Gold Portfolio 1970-1992
Reliant Scimitar 1964-1986
Riley 11/2 & 21/2 Litre Gold Portfolio 1945-1955
Rolls Royce Silver Cloud Gold Portfolio 1955-1965
Rolls Royce Silver Shadow 1965-1981
Rover P4 1949-1959
Rover P4 1955-1964
Rover 3 & 3.5 Litre Gold Portfolio 1958-1973
Rover 2000 + 2200 1963-1977
Rover 3500 1968-1977
Rover 3500 & Vitesse 1976-1986
Saab Sonett Collection No.1 1966-1974
Saab Turbo 1976-1983
Shelby Mustang Muscle Portfolio 1965-1970
Stubebaker Gold Portfolio 1947-1966
Stubebaker Hawks & Larks 1956-1963
Sunbeam Tiger & Alpine Gold Portfolio 1959-1967
Thunderbird 1955-1957
Thunderbird 1958-1963
Thunderbird 1964-1976
Toyota Land Cruiser 1956-1984
Toyota MR2 1984-1988
Triumph 2000. 2.5. 2500 1963-1977
Triumph GT6 1966-1974
Triumph Spitfire Gold Portfolio 1962-1980
Triumph Stag 1970-1980
Triumph Stag Collection No.1 1970-1984
Triumph TR2 & TR3 1952-60
Triumph TR4-TR5-TR250 1961-1968
Triumph TR6 Gold Portfolio 1969-1976
Triumph TR7 & TR8 1975-1982
Triumph Herald 1959-1971
Triumph Vitesse 1962-1971
TVR Gold Portfolio 1959-1990
Valiant 1960-1962
VW Beetle Collection No.1 1970-1982
VW Golf GTi 1976-1986
VW Karmann Ghia 1955-1982
VW Kubelwagen 1940-1975
VW Scirocco 1974-1981
VW Bus. Camper. Van 1954-1967
VW Bus. Camper. Van 1968-1979
VW Bus. Camper. Van 1979-1989
Volvo 120 1956-1970
Volvo 1800 Gold Portfolio 1960-1973

BROOKLANDS ROAD & TRACK SERIES

Road & Track on Alfa Romeo 1949-1963
Road & Track on Alfa Romeo 1964-1970
Road & Track on Alfa Romeo 1971-1976
Road & Track on Alfa Romeo 1977-1989
Road & Track on Aston Martin 1962-1990
Road & Track on Auburn Cord and Duesenburg 1952-1984
Road & Track on Audi & Auto Union 1952-1980
Road & Track on Audi 1980-1986
Road & Track on Austin Healey 1953-1970
Road & Track on BMW Cars 1966-1974
Road & Track on BMW Cars 1975-1978
Road & Track on BMW Cars 1979-1983
Road & Track on Cobra, Shelby & GT40 1962-1983
Road & Track on Corvette 1953-1967
Road & Track on Corvette 1968-1982
Road & Track on Corvette 1982-1986
Road & Track on Corvette 1986-1990
Road & Track on Datsun Z 1970-1983
Road & Track on Ferrari 1950-1968
Road & Track on Ferrari 1968-1974
Road & Track on Ferrari 1975-1981
Road & Track on Ferrari 1981-1984
Road & Track on Ferrari 1984-1988
Road & Track on Fiat Sports Cars 1968-1987
Road & Track on Jaguar 1950-1960
Road & Track on Jaguar 1961-1968
Road & Track on Jaguar 1968-1974
Road & Track on Jaguar 1974-1982
Road & Track on Jaguar 1983-1989
Road & Track on Lamborghini 1964-1985
Road & Track on Lotus 1972-1981
Road & Track on Maserati 1952-1974
Road & Track on Maserati 1975-1983
Road & Track on Mazda RX7 1978-1986
Road & Track on Mazda RX7 & MX5 Miata 1986-1991
Road & Track on Mercedes 1952-1962
Road & Track on Mercedes 1963-1970
Road & Track on Mercedes 1971-1979
Road & Track on Mercedes 1980-1987
Road & Track on MG Sports Cars 1949-1961
Road & Track on MG Sprots Cars 1962-1980
Road & Track on Mustang 1964-1977
Road & Track on Nissan 300-ZX & Turbo 1984-1989
Road & Track on Peugeot 1955-1986
Road & Track on Pontiac 1960-1983
Road & Track on Porsche 1961-1967
Road & Track on Porsche 1968-1971
Road & Track on Porsche 1972-1975
Road & Track on Porsche 1975-1978
Road & Track on Porsche 1979-1982
Road & Track on Porsche 1982-1985
Road & Track on Porsche 1985-1988
Road & Track on Rolls Royce & B'ley 1950-1965
Road & Track on Rolls Royce & B'ley 1966-1984
Road & Track on Saab 1955-1985
Road & Track on Toyota Sports & GT Cars 1966-1984
Road & Track on Triumph Sports Cars 1953-1967
Road & Track on Triumph Sports Cars 1967-1974
Road & Track on Triumph Sports Cars 1974-1982
Road & Track on Volkswagen 1951-1968
Road & Track on Volkswagen 1968-1978
Road & Track on Volkswagen 1978-1985
Road & Track on Volvo 1957-1974
Road & Track on Volvo 1975-1985
Road & Track - Henry Manney at Large and Abroad

BROOKLANDS CAR AND DRIVER SERIES

Car and Driver on BMW 1955-1977
Car and Driver on BMW 1977-1985
Car and Driver on Cobra, Shelby & Ford GT 40 1963-1984
Car and Driver on Corvette 1956-1967
Car and Driver on Corvette 1968-1977
Car and Driver on Corvette 1978-1982
Car and Driver on Corvette 1983-1988
Car and Driver on Datsun Z 1600 & 2000 1966-1984
Car and Driver on Ferrari 1955-1962
Car and Driver on Ferrari 1963-1975
Car and Driver on Ferrari 1976-1983
Car and Driver on Mopar 1956-1967
Car and Driver on Mopar 1968-1975
Car and Driver on Mustang 1964-1972
Car and Driver on Pontiac 1961-1975
Car and Driver on Porsche 1955-1962
Car and Driver on Porsche 1963-1970
Car and Driver on Porsche 1970-1976
Car and Driver on Porsche 1977-1981
Car and Driver on Porsche 1982-1986
Car and Driver on Saab 1956-1985
Car and Driver on Volvo 1955-1986

BROOKLANDS PRACTICAL CLASSICS SERIES

PC on Austin A40 Restoration
PC on Land Rover Restoration
PC on Metalworking in Restoration
PC on Midget/Sprite Restoration
PC on Mini Cooper Restoration
PC on MGB Restoration
PC on Morris Minor Restoration
PC on Sunbeam Rapier Restoration
PC on Triumph Herald/Vitesse
PC on Triumph Spitfire Restoration
PC on VW Beetle Restoration
PC on 1930s Car Restoration

BROOKLANDS HOT ROD 'MUSCLECAR & HI-PO ENGINE SERIES

Chevy 265 & 283
Chevy 302 & 327
Chevy 348 & 409
Chevy 350 & 400
Chevy 396 & 427
Chevy 454 thru 512
Chrysler Hemi
Chrysler 273, 318, 340 & 360
Chrysler 361, 383, 400, 413, 426, 440
Ford 289, 302, Boss 302 & 351W
Ford 351C & Boss 351
Ford Big Block

BROOKLANDS MILITARY VEHICLES SERIES

Allied Mil. Vehicles No.1 1942-1945
Allied Mil. Vehicles No.2 1941-1946
Dodge Mil. Vehicles Col. 1 1940-1945
Military Jeeps 1941-1945
Off Road Jeeps 1944-1971
Hail to the Jeep
Complete WW2 Military Jeep Manual
US Military Vehicles 1941-1945
US Army Military Vehicles WW2-TM9-2800

BROOKLANDS HOT ROD RESTORATION SERIES

Auto Restoration Tips & Techniques
Basic Bodywork Tips & Techniques
Basic Painting Tips & Techniques
Camaro Restoration Tips & Techniques
Chevrolet High Performance Tips & Techniques
Chevy-GMC Pickup Repair
Custom Painting Tips & Techniques
Engine Swapping Tips & Techniques
Ford Pickup Repair
How to Build a Street Rod
Mustang Restoration Tips & Techniques
Performance Tuning - Chevrolets of the '60s
Performance Tuning - Ford of the '60s
Performance Tuning - Mopars of the '60s
Performance Tuning - Pontiacs of the '60s

BROOKLANDS BOOKS

CONTENTS

BROOKLANDS BOOKS

ACKNOWLEDGEMENTS

About six months ago whilst driving on the freeway from Los Angeles to San Diego, I was impressed by a beautifully preserved yellow Karmann-Ghia coupé that leisurely overtook me, a few minutes later a black convertible slid by and this started me counting. In all some seventeen K-G were spotted that day. This subsequently inspired me to search through some four thousand magazines and the distilled essence of my reading will be found on the following pages.

Karmann-Ghia owners have not been well catered for with regard to books published about their cars and it is hoped that by making available these out of print articles once again it will generate interest in the marque and encourage owners to preserve and restore them for the future.

I am sure that K-G enthusiasts will wish to join with me in thanking the management of Autocar, Autosport, Car, Car & Driver, Cars Illustrated, Modern Motor, Road & Track, Road Test, Sports Car World and Wheels for allowing their interesting copyright articles to be included in this work.

R.M. Clarke

———————————

The above was written nearly four years ago in the winter of 1981 and happily since then a good new book by Jonathan Wood "VW and Karman-Ghia: A Collectors Guide" has made it into the bookstores. The marque today is in even greater demand than it was then and as a tribute to its present status we have added a ten story supplement to satisfy the needs of current owners and restorers.

Our thanks are extended to the publishers of Classic & Sportscar, Hot Rod, Motor Trend, Sports Car Graphic and the World Car Catalogue for allowing us to include their copyright articles.

R.M. Clarke

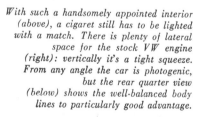

The GHIA-KARMANN
VOLKSWAGEN

photography: Poole

With such a handsomely appointed interior (above), a cigaret still has to be lighted with a match. There is plenty of lateral space for the stock VW engine (right): vertically it's a tight squeeze. From any angle the car is photogenic, but the rear quarter view (below) shows the well-balanced body lines to particularly good advantage.

As everybody knows by now, the most popular imported car in the U.S. by a wide margin is the Volkswagen. In the state of California, it now ranks 7th in sales, and demand is still well ahead of supply. But, like Detroit manufacturers, even Volkswagen has to look ahead; in fact, there are already a fussy few who quibble about the looks of the present Wolfsburg product. Accordingly, a Motorama-type feeler has been put forth to test public opinion on future styling changes for the VW sedan; it is a coupe designed by Ghia and built by Karmann (the German custom body firm which turns out the VW convertible), and which for brevity's sake we shall refer to as the VW-GK. Unlike most American "dream cars," however, this coupe is actually available for purchase providing the customer is well supplied with patience.

If looks are paramount, there is little doubt that the feeler will find fertile soil. The VW-GK, introduced at the European shows last fall and only recently on view here, has an almost universal appeal to the eye. It is, as the French would say, *une poupée vivante.* As can be seen from the accompanying photos, the car's Italian lines are low, beautifully balanced and ornament free. Whether intentional or not, there is very little suggestion that the engine is in the rear; the air-intake slots on the rear deck lid are the only reminder. Our test car, supplied by Allred Bros. of Glendale, came with an unusual two-tone paint job (metallic greenish-gold body and dark green top), but a wide choice of colors are available, most of which are quite a bit cheerier than those used on the standard sedan.

The interior of the car shows at once the touch of "custom" craftsmanship. Finish and attention to detail are excellent throughout. Two large, comfortable "contour" seats are provided for driver and passenger, and the seating position is far lower than in the sedan—almost, in fact, like a sports car. Fortunately the seats have a tremendous amount of fore-and-aft travel and can be adjusted to give full stretch even to the longest legs. If only the steering column were also adjustable! In spite of low seating, the top is proportionally even lower, and tall persons pay the usual penalty for sporty lines. Behind the front seats is a small flat bench with a vertical back, thinly padded and without springing. The back folds forward revealing a broad but shallow luggage compartment. For an occupant seated anywhere in the car, all-around visibility is unexcelled; the constant radius glass in the windshield and rear window contains an absolute minimum of distortion, and the four corner-posts are no more than an inch wide. The usual VW heating system is provided, but separate manual controls are added for each defrosting slot. The dash contains a 90 mph speedometer similar to but larger than that of the sedan, and there is a matching electric clock juxtaposed; indicator needle and hands are tapered for accurate reading.

Aside from the bodywork, all components of the VW-GK are from the standard VW sedan. As might be expected, therefore, the GK drives, feels, and sounds pretty much like the sedan. A trip ot the scales, however, indicated that it weighs about 120 lbs. more; even so, the weight distribution is identical—56.5% on the rear wheels. Underway to the testing strip, we found the noise level to be about the same as in the sedan. But the good aerodynamic lines of the GK make it far less susceptible to wind buffeting at highway speeds than its higher, more slab-sided cousin, and its somewhat lower center of gravity gives a more secure, stable feeling in tight spots. Unfortunately there is seldom a gain without a loss, and ventilation is something of a problem. There are no vent-panes in the wind-up windows, and the rear quarter-panes are fixed. Result: lower the windows even a little and your head is blasted; close them tight and you suffocate. The heater is no help since, unlike more complicated systems, it supplies only hot air. The point sounds minor but makes for chronic discomfort. *Continued on page* 14

Front compartment has the new-shaped gas tank, standard on all '56 VWs, which gives added storage space.

ROAD AND TRACK ROAD TEST NO. F-5-56

VOLKSWAGEN COUPE

SPECIFICATIONS

List price	$2475
Wheelbase	94.5 in.
Tread, front	50.8
rear	49.2
Tire size	5.60-15
Curb weight	1760
distribution	43.5/56.5
Test weight	2110
Engine	flat 4
Valves	pohv
Bore & stroke	3.03 x 2.52 in.
Displacement	1192 cc
Compression ratio	6.60
Horsepower	36
peaking speed	3700
equivalent mph	74.8
Torque, ft/lbs	56
peaking speed	2000
equivalent mph	40.4
Mph per 1000 rpm	20.2
Mph at 2500 fpm	120
Gear ratios (overall)	
4th	3.61
3rd	5.41
2nd	8.27
1st	15.8
R&T high gear performance factor	26.9

PERFORMANCE

Timed top speed	76
Max. speeds in gears—	
3rd (4500)	61
2nd (4500)	40
1st (4550)	21
Shift points from—	
3rd (4300)	58
2nd (4300)	38
1st (4150)	19
Mileage range	30/35 mpg

ACCELERATION

0-30 mph	6.9 secs.
0-40 mph	11.9 secs.
0-50 mph	18.2 secs.
0-60 mph	28.8 secs.
0-70 mph	49.2 secs.
Standing ¼ mile	23.6 secs.

TAPLEY READINGS

Gear	Lbs/ton	Mph	Grade
1st	400	15	20%
2nd	300	22	15%
3rd	190	32	10%
4th	115	37	6%

Total drag at 60 mph, 105 lbs

SPEEDO ERROR

Indicated	Actual
30 mph	29.9
40 mph	38.4
50 mph	46.7
60 mph	56.0
70 mph	64.9
83 mph	77.9

VOLKSWAGEN COUPE

Acceleration through the gears

ROAD and TRACK

LUXURY version of VW, designed by Ghia and built by Karmann, is popular in Europe, but this privately imported car is first here.

GHIA VW IS HERE!

German consul's Karmann-Ghia coupe was landed in Sydney recently

ENGINE is a completely standard VW unit—but lightness of special body allows top speed of 72 m.p.h. instead of 68.

INTERIOR is simple but impeccably finished. The coupe seats four, is 13ft. 7in. long, 5ft. 4in. wide, 4in. high. Mechanical specifications are those of normal Volkswagen, except that a stabiliser bar is added in front. Price is about 60 percent higher than the VW—nearly £1600 in Australia, if it's ever imported.

SLEEKNESS is the keynote of the Ghia-designed coachwork that clothes the familiar VW rear-engined chassis. It does not pretend to be "orthodox" and results in a most attractive two-seater coupé.

THE Volkswagen, in its bread-and-butter form, has had a phenomenal success. It is by no means an attractive car, either in appearance or performance, but its very ugliness has even added to its appeal, exactly as has happened with a certain French "people's car". Long life and an excellent service organization have been the selling features all along, and at a time when shoddy workmanship and non-existent after-sales service have been all too prevalent among other makes, this has been sufficient to ensure

JOHN BOLSTER TESTS

an ever-increasing number of satisfied customers.

Yet, there has been a demand, especially in America, for a good-looking version of the humble VW. Germans are thorough and painstaking engineers, but they are not an artistic race. So, Ghia, the master, has been employed to style the new model. Let me say, straight away, that the result is superb. The new body, designed by Ghia and built by Karmann, has a purity of line and a perfection of proportion that almost takes one's breath away.

This is not only a very lovely car, but it is a new artistic conception. For years, it has been conventional to worship the cult of the long bonnet, and some hideous vehicles have been applauded in consequence. It all stems from antique times, when the car with a long nose was assumed to have a powerful engine, and therefore to be excitingly fast. This has been the reason for the victories in *concours d'élégance* of ill-proportioned devices, in which the box containing the engine has occupied most of the chassis and the wealthy proprietor and his mink-clad companion have been banished to the infernal regions near the back axle.

The Karmann-Ghia body is one of the most beautiful ever built because it does not pretend to be what it is not. It

A Karmann-Ghia VW

Beauty and Roadholding are combined in this Luxury Volkswagen

admits, without labouring the point, that there is an engine at the rear, and then goes all out to look luxuriously aerodynamic, or aerodynamically luxurious, whichever you prefer. It is a roomy two-seater, with immense luggage space behind the seats. The luggage compartment can be turned into further seating space, but this is only useful to those who have a couple of legless children. The petrol tank and spare wheel virtually fill the front bonnet.

Mechanically, the new model does not differ from the family version. The well-known air-cooled "boxer" engine occupies the stern sheets, and drives forward to the four-speed gearbox that is one of the best features of the car. The rear suspension is independent by swing axles, and in front one finds trailing arms. The one mechanical difference is the use of an anti-roll bar in front, but this simple improvement transforms the handling of the car.

When I first took over the Karmann-Ghia I was frankly disappointed with it. In London's traffic it lacked "snap", for the new body adds appreciable weight. It seemed noisy, too, and even the delectable gearbox did not add greatly

to the performance, because of the limited revs to which the engine could usefully ascend.

Yet, out on the open road, this suddenly became a most desirable car! For one thing, the aerodynamic body had considerably augmented the maximum speed. The standard VW will only struggle up into the middle 60s, but this one passes 70 m.p.h. with ease, registering a mean of 72.5 m.p.h. Furthermore, it seems willing to run all day at such a speed. As with all the VWs I have tested, the speedometer was hilariously fast, hurrying round to the "80" mark on every possible occasion. Nevertheless, this car has an effortless high cruising speed that will wear down the opposition of more powerful vehicles. It is remarkably quiet as regards wind noises, and the sound of the mechanism seems to be left behind. Fast cruising is an effortless and quite economical business.

The improvement in riding and handling has to be experienced to be believed. Gone altogether is the typical VW oversteer, and one can drive fast on wet roads without that uneasy feeling that the tail is about to wag the dog. Gone,

LUGGAGE accommodation is extensive although reached from the interior. The floor of the compartment lifts to form the backrest of a very "occasional" seat.

POWER UNIT is the familiar 1,192 c.c. flat-four engine, the same as that which propels the less glamorous VW saloon. Accessibility is only reasonable but the unit is designed for easy removal.

AUTOSPORT, FEBRUARY 15, 1957

ITALO-GERMAN co-operation has combined art with engineering. As in the saloon, the fuel and spare wheel are carried under the front "bonnet".

too, is that choppy up and down movement at low speeds on bumpy lanes. The weight distribution has obviously been improved, and the centre of gravity has been lowered.

The result of all this is a car which covers the miles in a most effortless fashion, almost irrespective of road surface. A slight slackening of speed for gradient or curve is merely an excuse to use that central gear lever. A touch of third speed up to 60 m.p.h. soon restores the normal velocity. At low speeds in towns, one is only human if one enjoys the admiring glances and flattering comments of passers by. Even when driven hard, this Volkswagen will achieve a full 35 m.p.g., so its expensive appearance is allied with economy of operation.

The air-cooled engine starts well and warms up quickly. It is noisier than a conventional unit at town speeds, but I think that one would soon become used to the slightly more metallic note emitted by the finned cylinders. The heating and demisting system is very powerful and rapid in action, an unusual virtue among rear-engined cars. A "hot" smell was occasionally noticed, but this was probably only due to new paint.

The driving position, all-round visibility, and layout of the controls are all first-class, and there is ample room for the long-legged motorist. The gravity petrol tank has a reserve tap, but one nevertheless misses a fuel gauge. I have become accustomed to watching my petrol level on a journey, though I do admit that not all gauges are accurate.

Rear-engined cars usually excel in mechanical accessibility, but the VW is not particularly noteworthy in this respect. However, it is specifically designed for extremely rapid removal of the engine-transmission assembly, and it is best to carry out this operation before attempting major repairs. In any case, routine maintenance is easy, and this is all that is likely to be required for many thousands of miles.

One frequently hears rumours that the everyday Volkswagen is to be replaced by a new model. If this is so, the Karmann-Ghia coupé may well be an indication of the shape of things to come. Certainly it proves that the engineers at Wolfsburg can overcome the well-known handling and riding deficiencies whenever they wish. Thus, any new model may be an even greater menace to its competitors than the present well-tried machine.

It remains for me to thank Mr. Cargill for lending me his very beautiful car. I understand that he intends to find a little more power for those standing starts at the traffic lights, which should remove the only real defect. This is an expensive model, but its appearance and finish are certainly of the highest quality.

SPECIFICATION AND PERFORMANCE DATA

Car Tested: Volkswagen Karmann-Ghia 2-seater fixed-head coupé. Price £1,216 7s. including P.T.

Engine: Four-cylinders 77 mm. x 64 mm., 1.192 c.c. Air-cooled flat-four. Pushrod operated overhead valves. Compression ratio 6.6 to 1, 36 b.h.p. at 3,700 r.p.m. Solex downdraught carburetter. Coil and distributor ignition.

Transmission: Single dry plate clutch. Four-speed gearbox with synchromesh on upper three gears and central remote control, ratios 3.63, 5.45, 8.33, and 15.95 to 1. Spiral bevel final drive to swing axles.

Chassis: Platform-type chassis with backbone. Independent front suspension by trailing arms and laminated torsion bars. Anti-roll torsion bar. Worm and nut steering gear. Independent rear suspension by swing axles and torsion bars. Double acting telescopic dampers all round, 5.60-15 ins. tyres on disc wheels. Hydraulic brakes.

Equipment: 6-volt lighting and starting. Speedometer, clock, warning lamps, flashing indicators, heater and demister.

Dimensions: Wheelbase 7 ft. 10½ ins. Track (front) 4 ft. 3 ins. (rear) 4 ft. 1¼ ins. Overall length 13 ft. 7 ins. Width 5 ft. 4¼ ins. Turning circle 36 ft. Weight 16 cwt.

Performance: Maximum speed 72.5 m.p.h. Speeds in gears, 3rd 60 m.p.h., 2nd 38 m.p.h., 1st 18 m.p.h. Standing quarter-mile 24.2 secs. Acceleration, 0-30 m.p.h. 7.4 secs., 0-50 m.p.h. 23.4 secs., 0-60 m.p.h. 36.2 secs.

Fuel consumption: 35.2 m.p.g.

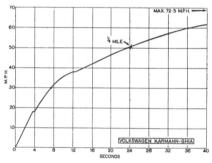

Acceleration Graph

ROAD TEST: **VW-OKRASA**

IF you want more performance from your German beetle, there are four routes to take—and they all cost money.

1. Install high-compression pistons, a reground camshaft, etc.
2. Buy the Okrasa Kit
3. Add a supercharger
4. Install a Porsche engine

Of these four alternatives, we have tested (briefly) several examples of the first possibility, and they give a definite improvement at the lowest cost, provided that you do the installation yourself. The acceleration time to 60 mph (after a speedometer correction) is about 24 to 25 seconds, and the top speed of a 1955 sedan (timed, of course) was 76/77 mph with the speedo at 83.

Item No. 2 is the primary subject of this test.

Item No. 3 (supercharging) offers the most performance gain per dollar spent. However, successful and satisfactory operation depends on two important factors. First, the air/fuel ratio must be checked out carefully, as too lean a mixture will produce overheating. Secondly, the driver must use his extra performance with proper re-

This shows the Okrasa kit installed in a VW sedan.

spect for an air-cooled engine. Cooling is dependent entirely on air flow, and the cooling fan must turn-up fast enough to keep a good blast of air going yet not turn so fast as to cavitate. In other words, with supercharging do not lug the engine, yet do not overspeed.

Bearing the above in mind, we have no objections to supercharging the VW, and our test work to date indicates a zero to 60 time of 16 to 17 seconds, a top speed of almost 90 mph with a sedan.

The limitations and possibilities of installing a Porsche engine, item No. 4, were thoroughly explored in our correspondence columns last year, and in the final analysis a Porsche engine is too expensive.

The subject of this test report is a VW-GK coupe, standard in every respect except that it was fitted with the "Okrasa Kit" manufactured in Germany. The cost of this kit is $249.50 plus installation and our test car was loaned to us by the importer, European Motor Products Inc., Box 668, Riverside, California.

Reserving a technical discussion on the kit itself for later, the test results show a startling and worthwhile improvement, to wit:

	STOCK '56	OKRASA
Top speed (cpe)	76.0	86.5
best run	77.9	87.5
Acceleration		
0-30	6.9	5.5
0-60	28.8	18.4
0-70	49.2	30.0
SS ¼	23.6	20.0

The data speaks for itself and in both cases refers to the "GK" or Ghia-Karmann coupe which, though 120 lbs heavier than a standard VW sedan, will normally out-perform the sedan in the higher speed ranges and in timed top speed.

Conservatively, the Okrasa Kit adds 10 mph to the honest top speed and give acceleration to 70 mph in about the same time as formerly required to attain 60 mph. More significant, the time to cover the standing-start ¼ mile is reduced by over 3 seconds, and the speed at the end is increased from about 56 mph to 62 mph. Calculations based on this performance indicate that if the standard VW develops an S.A.E. rating of 36 bhp at 3700 rpm, the Okrasa modification can reasonably claim 46 bhp at about 4200 rpm (Okrasa supply no dynamometer data).

Further calculations, using the Tapley meter lbs/ton data, prove that the low speed pulling power (or torque) is increased by 18%,

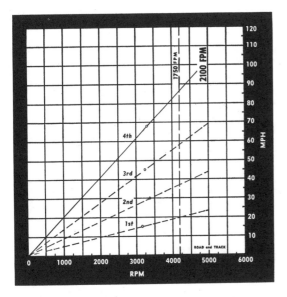

R & T ROAD TEST NO. 130

VW-OKRASA KIT

SPECIFICATIONS

List price (Okrasa kit)	$249.50
Wheelbase	94.5
Tread, f/r	50.8/49.2
Tire size	5.60-15
Curb weight, lbs	1760
distribution, %	43.5/56.5
Test weight	2080
Engine	flat-4, ohv
Bore & stroke	3.03x2.52
Displacement, cu in	72.7
cu cm	1192
Compression ratio	7.50
Horsepower (est.)	46
peaking speed	4200
equivalent mph	86.6
Torque, ft-lbs. (est.)	66
peaking speed	2100
equivalent mph	43.3
Gear ratios, overall	
4th	3.61
3rd	5.41
2nd	8.27
1st	15.8

CALCULATED DATA

Lbs/hp (test wt.)	43.3
Cu ft/ton mile	59.0
Engine revs/mile	2910
Piston travel, ft/mi	1220
Mph @ 2500 fpm	123

PERFORMANCE, Mph

Top speed, avg	86.5
best run	87.5
3rd (5000)	69.0
2nd (5000)	45.5
1st (5000)	23.3
see chart for shift points	
Mileage range	28/35 mpg

ACCELERATION, Secs.

0-30 mph	5.5
0-40 mph	8.6
0-50 mph	12.7
0-60 mph	18.4
0-70 mph	30.0
Standing start 1/4 mile	20.0

TAPLEY DATA, Lbs/ton

4th	165 @ 45 mph
3rd	295 @ 40 mph
2nd	400 @ 31 mph
1st	500 @ 19 mph
Total drag at 60 mph, 90 lbs.	

SPEEDO ERROR

Indicated	Actual
30 mph	30.3
40 mph	38.0
50 mph	47.6
60 mph	57.0
70 mph	66.5
80 mph	76.4
91 mph	87.5

VOLKSWAGEN G-K COUPES
Acceleration thru the gears

Puts a sting in the beetle

i.e., from 56 ft-lbs to 66 ft-lbs. Here, the gain can never equal the boost of a positive-displacement supercharger, but even an 18% improvement is well worth-while.

There is absolutely nothing about this kit which will give any outward clue as to the modifications—when riding with a driver who treats it very gently. It idles the same, sounds the same, feels the same. Yet anyone who owns and drives a normal VW will know instantly that something is different. You put it in 1st gear, tread on the gas, and you are past the red numeral I without realizing it. In fact, this brings up our one and only serious objection to the Okrasa Kit—it is much too easy and too tempting to over-rev. We ourselves over-rev the VW when getting the best possible acceleration data, but the red marks at 15, 30, and 45 mph are there for a purpose, and anyone who habitually exceeds these limits can expect trouble. Usually this takes the form of a broken crankshaft, and the only reason for such failures is excessive rpm, even though the actual break does not always occur at high speed. To facilitate analysis of the engine speed factor in the VW, we include here a chart of rpm vs. mph with circles showing the factory recommended limits and our own limit for occasional use (vertical dotted line), based on a piston speed of 1750 fpm.

So much for the performance and the one drawback. As shown here, the Okrasa Kit consists of special cylinder heads and related parts. Using special head castings accomplishes several important objectives: 1) It gives better cooling with more fins, 2) it increases compression ratio to 7.50 without the need for h.c. pistons, 3) it provides the necessary room for a much larger intake valve, and 4) it allows larger ports, not siamesed as in the standard heads.

Each of these four items is important and worthwhile: the improved cooling eliminates any sign of detonation which sometimes appears when 7.50 pistons alone are installed in a VW (but are OK with a reground camshaft); the higher compression gives both more power and better mpg; the power of an engine goes up in almost direct proportion to intake valve area increase; and elimination of the VW's siamesed intake ports makes twin carburetors a much more practical proposition. In summation, the Okrasa Kit tackles the basic problem correctly by beginning with the premise that new heads are the starting point. The rest of the changes, though important to the overall result, are incidental details. Although much more expensive than a supercharger, the Okrasa Kit makes the VW perform the way an 1192cc car (unblown) should. ●

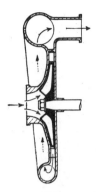
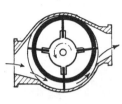

Three kinds of supercharger are (left to right), centrifugal, eccentric vane and three-lobe Roots

As mentioned earlier, they are geared accordingly. The Roots-type blower, on the other hand, produces relatively more "blow" at the lower end of its speed range than the others, its delivery curve almost flattens out for much of the operating speed range, and may fall away at extreme speeds due to rotor tip losses. The eccentric vane type has a single moving part, but it is out of balance and is thus speed limited. Roots blowers have two moving components, and are in balance.

There are few owners who would not be pleased to have a little more power for pick-up from the engines of their cars. Forgetting most of the considerations, the best way to obtain more torque low

L.P. MAG On The K.G. VW —Or Motosacoche's

Low-pressure Supercharging and its Application to the Karmann Ghia Volkswagen

SUPERCHARGERS are dogs given bad names—today quite undeservedly. Confusion of purpose, associated with the persistence of memories of long ago, put many car owners off supercharging who might well have become advocates. First, here are a few observations on the basic considerations, without reference to any particular type of "blower," before going on to discuss one kit and car—the MAG Karmann Ghia Volkswagen.

Superchargers are, of course, driven by the engine they boost, and the drive and speed ratio is selected to suit the work to be done. Special racing engines usually have superchargers working at very high pressures at fast running speeds; sometimes they are integral with the engine. Quite staggering power output can be achieved—485 b.h.p. from 1½-litres, for example. It is, perhaps, not surprising that engines so boosted have been known to "blow up" (not too literally).

The smart black Karmann Ghia Volkswagen is low overall yet has fair ground clearance and provides plenty of headroom. Mr. J. Scott Anthony, of Motosacoche, gives scale

On country roads outside Geneva the car handled in a pleasing and lively manner

These are not the kind of installations with which we are here concerned. The simple kinds of supercharger which can be added to a family or sports car engine are a different matter. But here again there must be a sub-division, depending upon the intention—it may be to increase maximum power output for modest competition work, or it may simply be to improve day-to-day acceleration, cruise performance, hill-climbing and flexibility, any gain in top-end performance being incidental. Performance in terms of favourable torque/r.p.m. characteristics is much more significant than high top speed to the motorist.

Most superchargers of the centrifugal and eccentric vane types produce manifold pressure in almost direct proportion to speed of rotation, and thus to engine r.p.m.

down would be to fit a bigger capacity engine. Another way is to force-feed the existing engine—that is, raise the initial pressure in the cylinders so that the combustion pressure is higher.

With mixture fed under positive pressure to the cylinders, one can forget the shortcomings of cheap manifold designs, non-registering ports and the other considerations of efficiency that John Davey has recently discussed in his Car Care and Improvement articles. Each cylinder can be expected to receive a full and equal share of well stirred-up petrol vapour/air mixture, regardless of the long, rough, curly pipes delivering it. For this reason, engines which have been lightly supercharged frequently run more smoothly, are more flexible and start well.

The owner need not fear greatly increased fuel consumption for a given performance, but if with the new-found power he drives a lot harder, then consumption will increase proportionately. Specific consumption with a supercharger at above zero manifold pressures is better than with the normally breathing engine. Cruising at negative manifold pressures, the consumption is only slightly worse. Putting up a better place-to-place average, but otherwise using speeds much as before the fitting of the blower, the increase in consumption should be less than 10 per cent.

A few engines of sporting character

The MAG supercharger is easily installed, and the complete rear power unit still occupies only a part of the available space

time to accelerate. The car corners very sweetly in answer to a light touch on the steering wheel. Like other VW models, it corners better with power on.

The particular car here reviewed is Swiss registered, and originally its performance, particularly on the many long gradients, left something to be desired. Our brief experience of the car was in the Geneva area, and road test equipment was not available there.

The owner, to whom we are much indebted for a loan of the car, is Monsieur Marcel Pithan, who is with B.P. He decided to fit a MAG type 70 supercharger kit, as developed for the Volkswagen by Motosacoche S.A. of Geneva. It is a neat Roots-type installation, which most owners could fit for themselves. A single belt drives the blower, from a rear-mounted pulley, at 1½-times engine speed. Normal manifold pressures are between 0 and 4 lb sq in.

have an unusual degree of valve overlap; it is as well to bear this in mind when considering supercharging because, particularly at the lower end of the speed range, it can be the cause of increased fuel consumption without gain in performance.

Worries about blue smoke trails and heavy oil consumption are almost entirely things of the past. A fraction more oil is needed to lubricate the supercharger, a little oil smoke may show in the exhaust when first starting up, but that is all. Engine idling usually needs to be set a fraction faster with a supercharger—at such low rotating speeds it still needs to be driven, but is giving nothing in return.

Turning now to the Karmann Ghia Volkswagen, here is a most attractive and pleasant-to-handle little car with two comfortable seats, luggage space, a sporting look but a rather flat performance. It has the same power (36 b.h.p. at 3,700 r.p.m.) as the standard VW and weighs 16 cwt in road trim. It cruises happily at 60 m.p.h. or more, but needs lots of

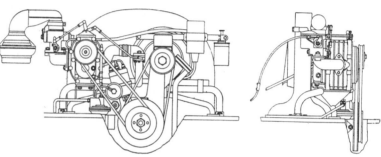

Installation drawings for the MAG supercharger in the Volkswagen

The makers stress that the increase in maximum speed is incidental, and that they normally choke down the carburettor to restrict top end performance. The intention, well realized, is to improve acceleration at lower speeds and to improve flexibility, without sacrifice of any sort.

A brief experience of the car on Swiss roads showed that it can now fairly be described as lively, particularly if the gears are used freely and, although the

lever movement is rather long, few changes are pleasanter to use. These remarks would apply equally well to the standard Volkswagen with the same supercharger installation.

A pleasant maximum speed for normal use in second gear is about 35 m.p.h., and in third the figure is nearly 60 m.p.h. Improvement in top speed, as already mentioned, is not considered to be important by MAG. This car was taken up to a speedometer reading of about 75 m.p.h., which was reached quickly and maintained without effort. There was still a little in hand. The instrument is probably a few m.p.h. fast.

What the driver appreciates is the flexibility now achieved, with considerably more liveliness. For the blown Karmann Ghia coupé, the top gear pickup at 40, 50 and 60 m.p.h. is considerably improved. The car's pleasant handling characteristics allowed its enhanced per-

Typical power curves for 1,100 c.c. four-cylinder engine. (1) standard tune 6.7 to 1 compression, 32 mm carburettor; (2) sports engine 7.6 to 1 compression, double throat 36 mm carburettor; (3) standard engine with MAG supercharging, but with top end throttle as recommended. The torque curves are in the same order as those for power

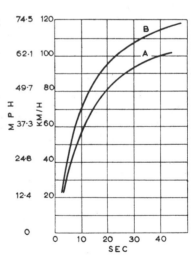

Maker's acceleration curves for the K.G. VW. (a) without supercharger, and (b) with

formance to be used freely on dry roads. In the course of a short fine weather drive the 30 front, 70 rear approximate weight distribution was not detectably different from the 42:58 of the ordinary Volkswagen.

One of the accompanying curves for a typical 1,100 c.c. engine indicates the improvement to be expected in power output—b.h.p. and torque—through the engine speed range. According to acceleration curves, also provided by Motosacoche, the standard model Volkswagen with blower is somewhat slower

BLOWN VOLKSWAGEN . . .

than the Karmann Ghia coupé so equipped, although their engines are identical and the latter is 1½ cwt heavier.

Certainly it is exceptionally clean aero-dynamically, and this might account for the relative improvement at speed.

Some comparisons are as follows:

	From Rest	
	10 sec	30 sec
VW standard	59 k.p.h.	90 k.p.h.
	36½ m.p.h.	56 m.p.h.
VW blown	67 k.p.h.	100 k.p.h.
	41½ m.p.h.	62 m.p.h.
K.G. VW	58 k.p.h.	93 k.p.h.
	36 m.p.h.	58 m.p.h.
K.G. VW blown	71½ k.p.h.	107 k.p.h.
	44½ m.p.h.	66½ m.p.h.

A small point of special interest in Switzerland is that the supercharged engine performance does not fall off to the same extent at high altitudes. According to whether a large or small air filter is fitted, the supercharger can either be heard as a faint, not unpleasant whine, or is scarcely detectable at all.

After sampling the special VW, an opportunity arose to drive for a few kilometres a Peugeot 403 with MAG supercharger kit. A word or two on this experience may be added as a postscript. In this case the improvement seemed more marked still. Not only had torque and flexibility at low speeds become so much

better that the built-in overdrive could be used like normal top, but acceleration of this spacious 1½-litre car, with the aid of the gears, becomes quite exhilarating.

Before the supercharger was fitted, this 403 accelerated from rest to 120 k.p.h. (74.5 m.p.h.) in 52sec through the gears. The figure is now quoted as 32sec. Top speed is nicely over 80 m.p.h. and the owner reported that he had found no adverse qualities at all.

The cost of MAG supercharger kits in Switzerland is moderate, elsewhere it varies considerably, reaching some $300 (well over twice the Swiss price) in America. Cars for which kits are available in addition to the above include Peugeot 203, Simca Aronde, Fiat 1100, Mercedes 180 and 220 and Volvo PV 444.

VW ROAD TEST

6▶

We have never been able to induce a stock VW sedan to break 70 mph for a timed 4-way average or beat 30 secs. from 0-to-60—and we have tested a good many. The GK, thanks to streamlining, will finally do the trick. Its best run was almost 78 mph (indicating 83) and its best 0-60, 27.7 secs. Because of the low specific output of the stock VW engine, finding an optimum rev limit for shifting is a problem. We used 4300 rpm on the GK for the most part, but found that in the 0-60 trials, winding up to an actual 60 in 3rd gave no appreciably better times than shifting to 4th just short of 60. Throughout the test the car had a good handling feeling about it—better so than the sedan—but this may well have been illusory in part, due to the lower seating position and greatly improved "grip" of the seats. Steering is unchanged at 2 2/3 turns lock-to-lock.

With only a few of the cars to be seen yet in the U.S., already the inevitable questions are beginning to arrive on our desk: "I have a VW-GK on order; what can I

Comfort is not a feature of the rear seat, but the back folds forward for luggage space.

do to improve its performance?" The general answers are, of course, the same as for the standard sedan. There are several makes of superchargers currently available for the VW, but in the GK there is a space problem: due to the fact that the rear deck

lid fits very closely over the engine there is no vertical room to install a blower. Only the Swiss MAG unit, which fits to one side, is readily adaptable. Undoubtedly the other makes will, however, soon have a modification of their blower suitable for installation on the GK. With a little extra push, the car's performance should be nearly as lively as a Porsche 1500, although the latter would still have a considerable advantage in smoothness and quietness.

When all is said and done, what does the VW-GK amount to? The overall performance improvement, we feel, is negligible. For nearly $1000 more than the sedan, then, the customer is acquiring a very pretty body. According to dealers' reports, that seems to be enough for a lot of people. Karmann is at present turning out between 300 to 400 cars per month; by fall the figure is expected to be 1000 per month. About half of this production is scheduled for the U.S., but the backlog of orders is already such that most dealers are quoting 2 to 4 years for delivery. *Some* body! ●

KARMANN—GHIA

(Continued from page 19)

the ownership of transport, and the possession of a skilfully-engineered motor car. Its price of approximately £1,600 is offset as well by the parts and service facilities offered by the Australian Volkswagen organisation.

Here are some brief mechanical details:

ENGINE: 4-cyl. air-cooled, o.h.v. 4-stroke, horizontally opposed. 30 b.h.p. at 3,400 r.p.m.; 36 S.A.E. horsepower at 3,700 r.p.m. Bore 77 m.m. x 64 m.m. stroke. Capacity 1192 c.c. Compression ratio 6.6 to 1. Piston speed, 1428 ft./min. Mechanical fuel pump, 6-volt ignition.

TRANSMISSION: Single dry-plate clutch, four speed gearbox with synchromesh on top three ratios. Ratios: 3.6, 5.3, 8.5, 15.8.

CHASSIS: Separate platform chassis.

SUSPENSION: Independent all around by trailing arms and torsion bars. Torsional stabiliser bar in front. Hydraulic telescopic shock absorbers.

BRAKES: 4-wheel hydraulic, leading and trailing shoes.

MEASUREMENTS: Weight, unladen, 1790 lb. Length, 163 in.; width, 64.3 in.; height, 52.4 in.; wheelbase, 94.5 in.; track, 51.4 in. (front) 49.2 in. (rear).

TYRES: 5.60 x 15 tubeless.   #

Graceful lines originally created by Italy's Ghia are combined with Germany's Karmann coachbuilding firm to form this handsome coupe.

KARMANN-GHIA:

Konnoisseur's Koupe

After several hundred miles behind the wheel of a Karmann-Ghia, ATHOL YEOMANS gives his impressions of one of these de luxe Volkswagens.

IT is a measure of the public interest in Volkswagen that more people will stop and peer into a Karmann-Ghia coupe *after* they discover the VW flash on the bonnet than before. Invariably, they will ask if it's the new Volkswagen, a sign of current interest in what Heinz Nordhoff, VW boss, describes as their "uniquely consistent styling policy".

For sure, the two most common questions in the minds of Australian motorists are:

What's the new Holden like? and When is VW going to bring out a new model?

And so, as the tally of beetles rises towards four million, and every day another three thousand make their exits from the factories in Germany and other parts of the world, the question gets louder:

When is VW going to make a new model?

Judging by the investment at present tied up in dies from which the bodies are stamped, Australian motorists may have a long wait. And the German branch of the family is naturally in no hurry to change the styling of something which is so much of a success that after twelve years of continual expansion of production facilities, demand for it still exceeds supply.

Only last year, daily output was raised by 600 vehicles per day—more than the entire daily output of Holdens — and in 1960 it will be raised again to 3,500 vehicles.

But if the beetle is going to be

KARMANN-GHIA: KONNOISSEUR'S KOUPE . . . *continued*

around for a long time yet, it will be something of a paradox, because although it will be the same, it will be different.

This is explained by the fact that the Volkswagenwerke have continually changed the mechanical specification of the car so that the car being sold now is a lot different from those of yesteryear. More powerful, more lavishly trimmed, more roadable, more comfortable— in every department gradual improvement has gone on so steadily that Nordhoff is quite justified in saying that Volkswagen have not lagged behind.

But Australia is certainly one of VW's most competitive markets, and, to hazard a guess, I'd say that pressure for a more modern appearance would be strongest here. However, although VW is currently number two in sales throughout Australia, the volume is not enough to warrant departure by the factory from their policy purely on our account.

Still, all good things come to an end, and sooner or later a re-styled VW will hit the deck. The question is:

There is some luggage space in the front of the Karmann-Ghia, but it is best suited for soft bags rather than bulky cases.

Rear occasional seat folds down to make additional space for luggage. Alternatively, a couple of children can be seated in the same spot.

Will it be a version of the Karmann-Ghia?

And the answer to that must be, regretfully; Highly unlikely.

The Karmann-Ghia is a different type of Kar, if you'll forgive me, from the Volkswagen, while any new model must be of the same sort.

The body of the Volkswagen must provide as much room as possible, for luggage and for passengers. A new or re-styled model must do something about those things in which the car is no longer competitive— more easily usable luggage accommodation, for instance. Although it would not, its fans would hope, pander to any motoring fashion, it could not afford to ignore persistent and established trends — right-around glass, the modern double-ended look.

Its components would have to satisfy production demands as well as those of styling. Panels would have to nest easily for transport, for instance. Curves must be capable of being stamped out in as few bites by the press as possible. To a degree, appearance and inside finish will be influenced by the final cost.

All this adds up to a lot different proposition to the Ghia.

In this we have a special series of car; the elegant coupe for two people, with some room for a fair bit of luggage or, for very short occasional trips, another person. The car, indeed, of universal appeal because of the fun it can give, but the car which the ordinary motorist can rarely afford to own, on the grounds on economic unsuitability if not on the actual price tag.

But on this question of price the Karmann-Ghia has one big advantage which special-series cars rarely enjoy — a chassis which is cheap to service, parts which are cheap to

Shape of the back makes the K-G look almost front engined. Bodywork is particularly well made and shapely.

Engine in the rear of the Karmann-Ghia is exactly the same as normal saloon. However, sound proofing is very good.

buy, service which is easy to obtain. This is a feature which seems to become more and more important every year.

To deal with the car itself. After several hundred miles in one of the handful of cars in the country, I still have to think very carefully about what it is that makes it such a car of character and appeal. Perhaps, most of all, it is the effortless way it goes about its job, its untiring gait, mile after mile, its ability to flatter the driver and make him feel a better pilot than he really is.

At 4 ft. 4½ in. it comes up to a point halfway between one's belt and tie knot, which is low. However, thanks to really wide doors, getting in and out is no more difficult than other low cars, although, of course, it's a bend and stretch compared to a high-standing big car.

Once inside, the comfort is excellent. Driver and passenger sit low in big, deep seats which have finger-light adjustments for reach and squab angle. Vision is commanding, and is as good to the rear as to the front. The controls fall easily to hand, with the reservation that an exceptionally long-legged driver, who has the seat well back, will have to reach for the gearshift. Another feature is that the pedals are offset to the left in relation to the driver's seat. The Karmann has a full-width body; the standard VW has not. On both cars the pedals are within the front wheel arches, but in the Kar-

KARMANN-GHIA: KONNOISSEUR'S KOUPE

Interior appointments in the Karmann-Ghia are very good. Seats are fabric and plastic. Like normal VW's, the steering wheel is now dished.

man the seat is further to the right. The result is that when placed straight ahead the left leg would fall on the accelerator pedal. Thus the offset is considerable, and although familiarity brings acceptance, drivers who change from one car to another won't like it.

Instrument layout is simple. There is a large clock, a large speedometer with a dead-steady needle, a very welcome fuel gauge (as well as the customary reserve tank) and warning lights for oil pressure and generator charge. On this car it also indicates the well-being of the cooling fan drive.

On the other side of the facia there is a glove box and a grab handle for the passenger. Two controls under the dash admit air to ducts at the base of the windscreen, and a similar duct defrosts the big rear window. These ducts also blow in cooling air at the front, but there is no ventilation around the feet. When heating is wanted warm air is blown in around the feet, through the windscreen ducts and through the rear window duct.

Other interesting features are the opening rear window panes, the excellent use of fabric and plastic in the interior, and the drop-down padded (but not upholstered) rear occasional seat squab which forms a largish luggage locker.

While a person used to loading a conventional boot will find stowing the Ghia awkward, it must be remembered that the car also comes in convertible style. When the op is down loading the car with luggage becomes even easier than with a conventional boot.

This convertible also removes the headroom problem for rear seat passengers. I found headroom in the front adequate.

On starting, the first thing one notices is the remarkable silence of the engine and transmission. This is partly due to the modifications recently introduced on all VW models, which include a reduction in fan speed, but is also due to the more liberal use of sound-deadening materials which can be used in a car costing £600 more than the standard saloon. The car is the quietest air-cooled vehicle I have driven, is at least as quiet as the average water-cooled car, but of course is not as silent as the more refined of the latter type.

However, the Karmann-Ghia has an advantage of many water-cooled cars because it has a shape which keeps wind noise low, and gearing which permits high speeds to be used without mechanical noise rising very much. One of the big attractions of the car is the quiet and effortless way it consumes miles at high speed without the intrusion of noise.

Matched with this feature are others. They include finger-light, precise steering; a first-class suspension and brakes with a bigger lining area than the VW saloon. Much has been said about the handling of VW's, but overall, taking bad conditions and good, high speeds and low, full loads and empty, straight roads and twisty ones, the roadability is good indeed. It has been improved noticeably with the recent modifications to the suspension. These include a lower pivot point where the axles leave the transmission, softer suspension in the in-

itial bump stages, a torsional stabiliser bar at the front. All this has reduced the oversteer appreciably.

Aided by a lower build, and the distribution of the extra weight of the KG (181 lb.) more towards the front wheels, it handles better still. On corners there is little roll and no tyre squeal until cornering forces become very high. In addition, the car can be flung about freely with every confidence. Avoiding a pot hole at high speed is a matter of a touch of the wheel. On dirt roads with a loose surface the Karmann-Ghia is much more sure-footed than cars which claim less oversteer, and on rough surfaces it gives a performance which has to be tried to be fully believed.

On all occasions the brakes proved up to their job, with light pedal pressures and progressive braking as the pressure was increased.

Although the top speed of the car is not high — fractionally better than 70 m.p.h. — and climbs on long grades call for frequent use of third and second gear if speed is to be maintained, the performance is wholly usable. In fact, on undulating country speeds consistently better than the top speed of the car can be obtained time and time again by using the run down the grades, with no protest from the car at all.

Passengers comment on the comfort of the ride, and the non-technical rarely seem to realise that they are motoring fast.

For night driving the lights are adequate for the top speed of the car, and headlight flashers are fitted to the end of the turn light indicator lever. These flashers can be used to warn overtaken vehicles by day as well as by night.

Mechanical accessibility is good, with the proviso that regular tappet adjustment is best done by the service station where the appropriate facilities are available. Apart from that, everything is in full view. Changing fuses on the test car in the dark was hampered by the local installation of radio gear.

Fuel consumption was 34 m.p.g. This included much full-throttle work, some country pottering and a leavening of traffic use.

Special mention must be made of the gearbox, which is proverbial for its easy, instant changes. A future modification could well be synchromesh on first gear, which would assist the average driver in city conditions. In all other conditions the gearbox is unexcelled, whether changes are snatched or taken leisurely.

The Karmann-Ghia has, in addition to the modifications mentioned here, all the other small changes made to standard VW saloons last year. Slight changes have also been made to the Karmann-Ghia exclusively, including the provision of windscreen washers as standard equipment.

All in all, the car is an elegant and practical coupe, of moderate but wholly usable performance, first-class roadability and which has the character and appeal which makes every mile as enjoyable as the first — and which makes the difference between *(Continued on page* 14 *)*

538   THE AUTOCAR, 7 APRIL 1961   645 BLN

1815

Volkswagen Karmann Ghia

There is little in the appearance of the VW Karmann Ghia to suggest that it is rear-engined, save perhaps for the absence of the conventional air intake in the nose. The two openings there are for interior ventilation

FROM time to time there emerges a car with a design so outstanding that it continues in production for a very long period without substantial change. Such a car is the Volkswagen, which was tested in its latest form with more powerful engine and all-synchromesh gearbox in October of last year. However, until now the special version of the Volkswagen—the Karmann Ghia coupé—has not been tested by this journal. This model utilizes all the main components of the saloon, the four-cylinder horizontally opposed 1,192 c.c. air-cooled engine, combined with gearbox and final drive at the rear of the car, transverse torsion bar suspension units, and rack-and-pinion steering.

The platform-type chassis, however, is wider to suit the ow and shapely Karmann Ghia body and it is this body shape which resulted in a higher performance than achieved with the saloon, in spite of a weight increase of nearly 1 cwt. The only significant difference in the engine compared with that in the standard saloon is that a ball check valve in the carburettor, which weakens the mixture at near maximum r.p.m., has been removed, enabling slightly higher engine speeds to be reached.

It was at medium to high road speeds that the increased performance was most evident. For example, 40-60 m.p.h. in top gear required 19·5sec, whereas the time for the saloon was 25·2sec. Similarly, when accelerating from rest to 60 m.p.h., the corresponding figures are 26·5 and 32·1sec, and the mean maximum speed of a Karmann Ghia was 77 m.p.h.—an increase of 5 m.p.h. over that of the saloon. With the VW engine the maximum speed is also the cruising speed, given suitable roads, for it is lightly stressed and is designed to run continuously at full throttle. Normally when cruising, however, the accelerator can be eased back and the Karmann Ghia will cruise at a true 70 m.p.h. on quite a small throttle opening with benefit to fuel consumption.

Downhill speed can be built up appreciably above the maximum obtainable on the level, and in this way a true 80 m.p.h. can be sustained at times. The engine can be heard plainly at all times but it is not obtrusive during fast cruising. It is more audible with the windows open and is muffled if the rear seat squab is in the raised position.

Performance is lively if full use is made of the excellent gearbox with its well-spaced intermediate ratios. Second and third, with maxima of 42 and 69 m.p.h. respectively, are very useful for overtaking, and the synchromesh is so effective that it can never be beaten, no matter how fast the lever is moved. With synchromesh also provided on first gear the task of driving in very heavy traffic is eased considerably, although it was a pity that the initial movement of the accelerator produced a rather jerky take-up. The central lever is well placed for reach and has a light precise movement between gears, accompanied by a small mechanical noise from the linkage as each gear position is reached. Reverse is easily engaged after first pressing down on the knob, an action which prevents this gear from being selected unintentionally.

There is much room for improvement in the arrangement of the pedals, which is inferior to that in the standard saloon.

Spare wheel and further luggage space in the nose. Ahead of the wheel are the jack, tools and spare fan belt and there is also stowage space for soft luggage. Beneath the platform is an 8·8 gallon fuel tank

The organ pedal accelerator is too near the horizontal, so that at full throttle the foot is in an uncomfortable position. The angle of the pads of brake and clutch pedals, when released, is beyond the vertical in the direction of the driver, so that his angle of attack on the pedals is awkward. Unless the driver wears shoes with narrow soles there is some difficulty in resting the left foot between the clutch pedal and the central floor member. Driver and passenger are placed farther apart than in the standard saloon, so that the driver sits diagonally with legs converging towards the centre of the toe board.

Requiring full depression for complete disengagement, the clutch has a light, smooth action and there was no slip when an easy restart was made on a 1-in-3 hill. The excellently placed handbrake, too, held the car without difficulty on the same gradient with the car facing up or down the hill. This was more than a parking brake, for it was effective with the car in motion and so placed that very good leverage could be applied.

Good Fuel Economy

In spite of an increase in compression ratio to 7 to 1 on all Volkswagen engines, regular grades of fuel still can be used, an important saving when large annual mileages are covered. On the Karmann Ghia there was a little pinking only when the engine was allowed to labour. During 1,113 miles covered by the test a fuel consumption of 31·2 m.p.g. was recorded and only when cruising continuously at near maximum speed did the figure fall below 30 m.p.g. In addition to the consumption at constant speeds, given in the performance table, this was also measured during a flat-out lap at an average speed of 77·6 m.p.h. on a banked circuit, when the extremely good figure of 27·6 m.p.g. was obtained. During brisk main road driving, or on twisty cross-country routes, figures of around 35 m.p.g. were recorded, and leisurely driving gave 40 to 44 m.p.g. according to road conditions, the car being fully laden.

With a tank capacity of 8·8 gallons, about 260 miles could be covered before having to run on the 1·1-gallon reserve, which is obtained by moving a lever on the toe board. The fuel filler, under the front boot lid, is of large diameter so that the tank can be filled rapidly without air locks. A contents gauge is fitted on the facia. At times, petrol fumes were noticed inside the car. An automatic choke is fitted and starting was instantaneous from cold, followed by un-faltering progress when driving during warming up. Hot starting, though usually good, sometimes required throttle manipulation.

Overall height of this coupe is only a little over 4ft 4in. so that entering and leaving the car is not a very easy matter in spite of the wide doors. The lower edges of the screen and windows are high in relation to the occupants; it is not possible for a driver of average height to see the left front

Seat backrests are adjustable for rake in three positions. Controls beneath the facia are the front boot lid release (beside the column) and levers to admit unheated air. On the toe-board above the dipswitch is the fuel reserve lever. Below: Children can be accommodated on the sponge rubber rear seat. There is storage space beneath the cushion

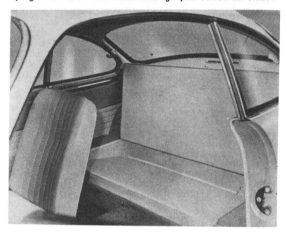

With the rear seat backrest lowered, it forms a carpeted platform for extra luggage. On the parcels shelf is a rear window demister slot

wing, although the low curved bonnet allows a good view of the road. Reversing is not as easy as it might be, because the driver is unable to see the rear extremities of the car. The wiper blades park above the lower edge of the screen and interfere somewhat with the vision on the nearside.

A good view to the rear is given by the interior mirror, and all pillars are very slender, providing good all round visibility and a light airy feeling inside the car. The steering wheel position and angle are well chosen, the driver assuming a comfortable relaxed position with his arms supported by his thumbs on the spokes of the wheel.

Steering is light and positive at all speeds, with hardly a trace of road reaction. The Karmann Ghia coupé retains the oversteering behaviour of the saloon but to a smaller extent. When cornering at moderate speeds it is necessary to feed back the wheel a little to hold the desired line, and when reaching the limit of adhesion the rear wheels broke away, but not in a vicious fashion. In fact this could be controlled very readily by applying opposite steering lock. In a long, sweeping bend it is not easy to make a single correction but instead several "bites" have to be taken. There is some tyre squeal during hard cornering.

With two persons and luggage aboard, rear wheel camber is about zero and there was little tendency for "jacking up" of the rear of the car during very hard cornering. Handling is improved by raising tyre pressures to those recommended for fast driving. Cornering is more stable if the throttle is held open, for lifting the accelerator on a bend at once causes the nose to tuck in.

Generally, there is good insulation from road noise,

Volkswagen... Karmann Ghia

Although the Karmann Ghia coupé weighs nearly 1 cwt more than the VW saloon, its lower height and smoother shape provide better performance at medium and high speeds

although a deep pothole can produce a loud thump. Springing is very well damped and the ride is comfortable on poor surfaces. A stretch of very bad Belgian pavé was taken at 35 m.p.h., which proved to be a comfortable speed with the car always under full control, neither slithering nor losing its line when cornering. The gear lever did not jump out of gear, although the facia locker lid would not remain shut and the sun visors dropped down. The car performed well also on a deeply corrugated surface, on which about 44 m.p.h. proved to be the most comfortable speed.

There was no opportunity during the test to assess the handling on wet roads but the adhesion of the German Dunlop B7 tyres appeared to be very high.

Having plenty of capacity for a car of this size, the brakes provided ample stopping power for all conditions encountered during the test, and they did not show signs of fade. When applied from high speed there was some roughness. Pedal loads demanded were rather large, 110lb being required to give the maximum retardation of 0·92g. There was a marked reluctance for the wheels to lock on a dry surface, and the brakes gave real confidence when applied in emergency stops.

An adequate area of the screen is cleared by the wipers and their motor is very quiet. Pulling out the wiper control switch operates the screen washer, which is a standard fitting.

Headlamps give sufficient illumination for the car's performance, but the position for the foot-operated dipswitch could be improved. It is set too high on the toe-board and is difficult to reach, so that the left foot must be passed around the clutch pedal. During a night run on a winding road the awkwardness of this dipswitch position was very obvious, when much gear-changing was called for.

For flashing the headlamps a switch is mounted on the underside of the lever for the turn indicators, this control being on the left of the steering column. Headlamps may be flashed in daytime without the side lamps being on, and if the switch is held the flashing is continual and automatic. In one respect this is a disadvantage, as it prevents the driver from holding the lamps on, as he may wish to do when signalling. All fuses are neatly grouped on the underside of the facia. Instrument illumination is varied for intensity by rotating the knob of the main lamp switch. There is some reflection in the screen at night from the steering wheel unless the instrument lighting is adjusted very low. Matching the speedometer is an unusually large electric clock which was very accurate.

Air Heating Arrangements

It would be an advantage with this air-cooled engine to have an oil temperature gauge. There is no indication of overheating unless the belt which drives the generator and cooling fan fails, in which case the lack of charge from the generator is shown by the usual warning lamp coming on. The Karmann Ghia has an efficient heating and ventilating system. Air warmed by the engine is admitted to the interior through openings at floor level, at the windscreen for demisting and defrosting, and at the rear window to keep this clear from condensation. The amount of heat is regulated by a sensitive control beside the seat on the central floor member. In addition, unheated air entering through two openings in the nose of the car can be mixed with air from the heating system by moving two small levers beneath the right of the facia, this unheated air entering through the windscreen demisting ducts. The amount of interior heating is reduced at low engine speeds.

Finish of both the interior and exterior of this coupé is of a high standard, and detail fittings are practical and well made. For example, there is an excellent over-centre catch for the hinged side windows, and when open these promote good circulation of air in warm weather. All seats are trimmed in good quality plastic, which was, however, such a light colour on the car tested that it showed dirt very quickly. Front seats are well shaped for good support, particularly against cornering forces, and their backrests are quickly adjustable in three positions. The fore-and-aft adjustment for the seats could have been easier to operate, but ample range was provided. Door windows were not up to the standard of the remainder of the car, as their winders were noisy.

Luggage space is very generous for a car of this size, there being a large carpeted space above the transmission;

The engine compartment remains very clean and all auxiliaries are readily accessible. There is provision for pre-heating the air for the carburettor intake; a flap-valve cuts this off as engine temperature rises

when the backrest of the occasional rear seat is folded flat, extra luggage can be carried here. A ledge prevents a suitcase from shifting forward during braking. There is further room for luggage in a shallow space above the fuel tank in the nose and yet more for soft luggage ahead of the spare wheel, where the tools and jack are also housed. There is a good and comprehensive set of hand tools and a spare driving belt for the cooling fan. Both engine compartment and front luggage boot lids are released from inside the car, the rear release being on the left of the rear seat. This would be better placed on the other side.

There are no restrictions on speed for running-in a VW except for the normal recommendations for maxima in gears which apply at any time. Ten chassis lubrication points require attention every 1,500 miles and three at 3,000-mile intervals.

In spite of its high price in this country, the VW Karmann Ghia is coveted because it is clearly a very stylish car and in most ways it is pleasing and untiring to drive. For touring it possesses two important attributes, economy during high-speed cruising and good luggage accommodation for its size.

VOLKSWAGEN KARMANN GHIA

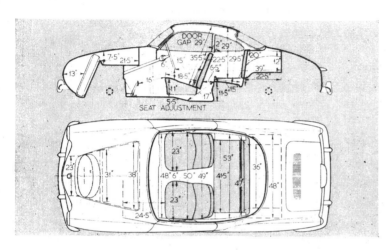

DOOR GAP 29"

SEAT ADJUSTMENT

Scale ¼in. to 1ft. Driving seat in central position. Cushions uncompressed.

DATA

PRICE (basic), with coupé body, £843 10s.
British purchase tax, £352 11s 8d.
Total (in Great Britain), £1,196 1s 8d.

ENGINE: Capacity, 1,192 c.c. (72·74 cu. in.).
Number of cylinders, 4, horizontally opposed, air-cooled.
Bore and stroke, 77 × 64mm (3·03 × 2·52in.).
Valve gear, overhead, pushrods and rockers.
Compression ratio, 7·0 to 1.
B.h.p. (net), 34 at 3,900 r.p.m. (Net b.h.p. per ton laden 36·4).
Torque, 61 lb ft at 2,000 r.p.m.
M.p.h. per 1,000 r.p.m. in top gear, 18·6.

WEIGHT: (with 5 gal fuel), 15·6 cwt (1,753 lb).
Weight distribution (per cent): F, 40·9; R, 59·1.
Laden as tested, 18·6 cwt (2,089 lb).
Lb per c.c. (laden), 1.75.

BRAKES: Type, Lockheed hydraulic.
Drum dimensions: F, 9·06 in. diameter; 1·57in. wide. R, 9·06in. diameter; 1·18in. wide.
Total swept area, 157 sq. in. (168 sq. in. per ton laden).

TYRES: 5·60—15in. Dunlop B7.
Pressures (p.s.i.): F, 16; R, 20 (normal). F, 17; R, 23 (fast driving).

TANK CAPACITY: 8·8 Imperial gallons, including 1·1 gallons reserve.
Oil sump, 4·4 pints.

DIMENSIONS: Wheelbase, 7ft. 10·5in.
Track: F, 4ft 3·4in.; R, 4ft 2·7in.
Length (overall), 13ft 7in.
Width, 5ft 4·3in.
Height, 4ft 4·4in.
Ground clearance, 6in.

ELECTRICAL SYSTEM: 6-volt; 66 ampère-hour battery.
Headlamps, 45–40 watt bulbs.

SUSPENSION: Front, trailing links and laminated torsion bars, telescopic hydraulic dampers, anti-roll bar.
Rear, independent, swing axles, round section torsion bars, telescopic hydraulic dampers.

PERFORMANCE

ACCELERATION TIMES:

Speed range, Gear Ratios and Time in Sec.

m.p.h.	3·89 to 1	5·78 to 1	9·01 to 1	16·63 to 1
10–30	—	9·1	5·7	—
20–40	14·4	8·9	7·2	—
30–50	16·0	10·8	—	—
40–60	19·5	15·2	—	—
50–70	28·4	—	—	—

From rest through gears to:

30 m.p.h.	..	6·6 sec
40 ,,	..	10·9 ,,
50 ,,	..	17·1 ,,
60 ,,	..	26·5 ,,

Standing quarter mile 23·0 sec.

MAXIMUM SPEEDS ON GEARS:

Gear			m.p.h.	k.p.h.
Top	..	(mean)	77·0	123·9
		(best)	78·0	125·6
3rd	..	..	69	111
2nd	..	..	42	68
1st	..	..	23	37

TRACTIVE EFFORT (by Tapley meter):

			Pull (lb per ton)	Equivalent gradient
Top	..	..	170	1 in 13·1
Third	..	..	255	1 in 8·7
Second	..	..	395	1 in 5·6

BRAKES (at 30 m.p.h. in neutral):

Pedal load in lb	Retardation	Equiv. stopping distance in ft
25	0·17g	177
50	0·34g	89
75	0·56g	54
100	0·85g	35
110	0·92g	32·8

FUEL CONSUMPTION (at steady speeds):

	Top gear
30 m.p.h.	56·3 m.p.g.
40 ,,	51·3 ,,
50 ,,	45·4 ,,
60 ,,	39·2 ,,
70 ,,	31·5 ,,

Overall fuel consumption for 1,113 miles, 31·2 m.p.g. (9·1 litres per 100 km).
Approximate normal range 28–40 m.p.g. (10·1–6·4 litres per 100 km).
Fuel: Regular grades.

TEST CONDITIONS: Weather: Dry, 0–5 m.p.h. wind.
Air temperature, 59 deg. F.

STEERING: Turning circle:
Between kerbs: L, 34ft 8in.; R, 32ft 1in.
Between walls: L, 36ft 7in.; R, 34ft.
Turns of steering wheel from lock to lock, 2·75.

SPEEDOMETER CORRECTION: m.p.h.

Car speedometer	..	..	..	10	20	30	40	50	60	70	80
True speed	..	..	..	10	19	28	37	45·5	55	64	74·5

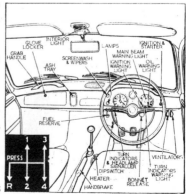

The works department.

VW LIFTS THE VEIL ON KARMANN

S.C.W. FULL ROAD TEST

With four more horses, the 1961 Ghia-styled version of the German People's car is near sports car class.

IN Australia there still are not many Karmann Ghias using our roads to delight the tired eyes of enthusiasts who really appreciate automobile styling. The reason is certainly the price — £1598 including tax. But there is not one among the 100,000 buyers of the ordinary VW who would

not chance the cramping of his rear seat passengers to drive one of these styling masterpieces.

SPORTS CAR WORLD tested the first Karmann Ghia back in May, 1960, though a handful had reached our shores earlier as the transport for those lucky members of the Consular Corps.

The 1961 KG, now boasting 40 brake horsepower instead of 36, seems a completely new car on performance. All of the former model's vices have been lost with this extra surge of power. In fact the newcomer, which along with the VW policy of only very, very minor changes, cannot be picked from outside, seems a gear better all round.

Addition last year of a stabiliser bar to the VW's certainly helped handling. But those four horses have made things even better, assisted of course by the low centre of gravity of the KG, with added weight, width and length, which is more than compensated for by its aerodynamic design.

If you want an honest but not outstanding performer (not forgetting it has an engine capacity of only 1192 cc) to attract attention, then the KG could be the solution. During our two days' test it attracted more attention than the Porsche 1600 Super.

The KG is still so uncommon that the average man in the street does not recognise it. First comment, from a family man, was: "Does it go any better than the ordinary VW?" The answer is 10 mph when you really have it wound up. Then a motor engineer asked if the KG was a Porsche, after which a young lady described it as a Renault Floride.

Our KG was the personal car of Lanock Motors Service Director Mr Bruce Fraser and had 7000 miles on the clock. Though VW's don't need any running in, it is accepted that they do run freer with some miles under the belt. This ruled out the brand of new vehicles of the firm's sales executives, Doug Donaldson and Bruce Gartrell.

The test KG was black with white roof, which brought home to me how hard black vehicles

are to keep clean, though they do look outstanding after you have spent half an hour on them.

Though quieter than early VW models, the KG still gives a lusty air-cooled sound as you rev up in the lower gears. Changes were easy and the synchromesh on the first gear exceedingly handy.

Only interior change in the KG was the shifting of the bonnet catch button from the passenger's side to next to the steering wheel.

The Hella headlights with European style cut-off for low beam proved good on the open highway. A thermostat gives a range of instrument lighting to suit individual tastes.

Most prominent dial, right in front of the driver, was the speedo, which lacks those red marks on the normal VW as a guide when to change gears up or down. Just as large on the left is a clock, an electric one which proved to keep perfect time. But what a waste. A rev counter here would prove much more satisfactory, giving an added sports car effect, as well as not tempting drivers to speed as they are ever reminded they are late for that vital appointment.

In the centre of these two dials is (hooray) a fuel gauge, divided into quarters for the 8.8 gallons tank, which includes 1.1 gallons operated off a reserve switch.

This gauge is accurate, but badly marked. The reserve is at the end of the last quarter, instead of halfway through it, giving you the false impression that you are not likely to reach the last ounce for some time.

But a loss of power, in our case in heavy traffic on the Princes Highway, soon gives you the message. and you switch to reserve. I found I had to stop to turn the tap and so found it hard to get back into traffic.

Our test car had a push button radio, cigarette lighter in the dashboard as on the older models and a small glove box. The headlights are on a pull out button with two stages, while another button switches on the wipers when you turn it round, or squirts water on to the screen each time you pull it out.

VW LIFFTS THE VEIL ON KARMANN Cont.

The KG convertible, available to special order only.

Bigger (though still tiny) boot with re-arranged petrol tank.

Wide opening door. The bonnet catch is now on the driver's side.

With switches to give fresh air ventilation to either driver or passenger, plus a good working heater, the KG has many safety features lacking in the real sports cars. Backrest of the seats can be adjusted to five positions to ease fatigue.

Each door has a wide pocket to accommodate maps and other extras. The self cancelling trafficator switch is on the left of the steering wheel. Pushing it in and out flicks the headlights in Continental style.

Room inside the KG is not extravagant, though sufficient. With the seat pushed fully back I could not quite use the straight arm technique (I'm 5 ft 10 in tall). The back casual seat is definitely not for adults, but for two children (under 12 years). This seat is comfortably made of foam rubber, but leg room is missing and tall persons are inclined to bump their heads when going over bad roads.

On cool nights all passengers, front and back, commented on the draughtiness of the KG with the driver's window down. The wind misses the driver and curls round the back of their necks. This was not lessened even with the back quarter windows open or the heater on.

Early in the test my clumsy right foot was inclined to stamp on both the accelerator and brake pedals at the one time, giving you quite an exhilarating effect as you want to slow down, but really go faster. This was due to the close spacing of the pedals, which I am told is the same in all VW's. But I must admit it never happened after I really got to know the KG.

Forgetting the color scheme, the KG is quick to clean with its flowing lines and high-class finish. But the chrome air vents in the nose come loose easily and could be made to rattle when handled.

The number plate resting on the front bumper appears vulnerable to traffic scratches, but such is not the case. The overriders are perfectly placed to ward off blows, except by a vehicle with a pointed back which could intrude between them.

Tyres were German Dunlop which did not squeal, even under tortuous cornering. Pressures are always interesting with rear engine vehicles, due to their greater weight on the back wheels. In the KG handbook owners are advised to have 16 lb front and 20 lb back for one or two passengers, or for high speeds or fully loaded, 17/23. But inside the glove box an extremely handy sticker advises 17/24 at all times. So the VW boys will have to get together on this one.

As usual I found cars handle better with a little extra air than advised by the makers. Admittedly the ride is harder over bumps and not so likely to sell cars. But handling on the highways at fast speed is assisted. Finally I settled for 20/25, which gave perfect conditions.   #

S.C.W. TEST No 29

VOLKSWAGEN KARMANN GHIA
(40 BHP)

PERFORMANCE

(FIGURES FOR 36 BHP MODEL are given in brackets)

TOP SPEED:

Two-way average	82.9	(77.6)
Fastest one way	84.9	(79.0)

ACCELERATION:

(test limit	3600 rpm)	
Through gears:		
0-30 mph	6.4	(6.3)
0-40 mph	10.2	(10.3)
0-50 mph	16.6	(17.4)
0-60 mph	24.1	(27.7)
0-70 mph	34.3	
Standing quarter-mile	21.2	(23.3)
Speed at end of quarter	59	(58)

MAXIMUM SPEEDS IN GEARS (INDICATED)

(at 3600 rpm)

I	29
II	52
III	74

SPEEDOMETER ERROR:

30 mph indicated — actual 32 mph
40 mph indicated — actual 41 mph
50 mph indicated — actual 50 mph
60 mph indicated — actual 61 mph
70 mph indicated — actual 71 mph

TAPLEY DATA:

MAXIMUM PULL IN GEARS:

I	550 lb/ton
II	450 lb/ton
III	340 lb/ton
IV	230 lb/ton

ROLLING RESISTANCE

70 mph	60 lb/ton
60 mph	40 lb/ton
50 mph	35 lb/ton
40 mph	20 lb/ton
30 mph	10 lb/ton

BRAKING:

From 20 mph: 15 ft 7 in (excellent)
Fade: Nil.

CALCULATED DATA:

Weight as tested (2 men)	18¾ cwt.	
Max bhp — 36 nett at 3600 rpm (30 at 3400)		
Max torque — 61 lb/ft at 2000 rpm (56).		
Lb/hp (nett)	69.3	
Mph/1000 rpm top gear	18.7	(20.0)
Mph at 2500 ft/min piston speed top gear		(120)
Cub./cm./lb ft torque		(24.8)
BHP/litre		(25.2)
BHP/ton as tested		(32.3)
Brake lining area/ton as tested		(86.8)
Piston speed at max. bhp		(1414.4)

SPECIFICATIONS

PRICE: £1598 (inc tax)

ENGINE:

Type: 4 cylinders, horizontally opposed, air cooled
Valves: OHV push rods
Capacity: 1192 cc.
Bore and Stroke: 77 x 64 mm
Piston area: 28.8 sq in.
Compression ratio: 7.0 to 1 (6.6)
Carburettor: Solex 28 pict.
Fuel Pump: Mechanical
Timing control: Vacuum.
Spark plugs: Bosch 175.
Max. power: 40 bhp at 3600.

CHASSIS:

Type: Tubular centre section, forked at rear; welded
    on platform.
Steering: VW worm gear.
Wheelbase: 7 ft. 10½ in.
Track, front: 4 ft. 3½ in.
    rear: 4 ft 2¾ in.
Suspension: front—independent by trailing arms, lamin-
    ated torsion bars and anti-roll torsion bar.
    rear: Independent by divided axle and laminated torsion
    bars.
Shock absorbers: Telescopic.
Fuel tank: 8.8 gal., inc. 1.1 gal. res.
Tyres: German Dunlop.
Wheels: Bolt-on steel disc.
Brakes: Type—Drum.
Operation: Hydraulic.
Lining: 9.06 in. diameter.

GEARBOX:

Four-speed (all synchro).
Clutch: Fichtel and Sachs, SDP.

RATIOS:

I	3.89
II	5.76
III	9.01
IV	16.6

GENERAL:

Length overall: 13 ft 7 in
Height: 4 ft 4½ in.
Width: 5 ft 4½ in.
Ground clearance: 6 in.
Turning circle: 36 ft.
Test weather: Fine, no wind.

All test runs made on dry, bitumen-bound gravel road
with driver and one passenger aboard. High-speed runs
made with windows up. All times averaged from several
runs in opposite directions, using a corrected
speedometer.

● Once upon a time there were two Volkswagens. The year was 1949 and the Americans who saw them shook their heads, laughed and bought domestic cars. Today over half a million of those same citizens are riding in cars which look virtually identical to the ones they once scorned. From a mere trickle, Volkswagen sales have developed near flood-tide proportions that almost boggle the imagination.

As you read this, Volkswagen of America will be putting the finishing touches on its $2.5 million headquarters located on an 18-acre site on the New

Jersey Palisades opposite New York City. At the same time, VW will have made another sale, boasting the 1961 total further over 200,000 units.

TOO BUSY SELLING VWS

Volkswagen of America's Carl H. Haan freely admits, "we were too busy selling Volkswagens" to do much with the sleek Karmann-Ghia coupes and convertibles. Yet they have won an admirable public acceptance with a total of over 9300 sold in 1961, compared to 2452 in 1956. The picture will change in

1962. Not that VOA won't be "busy selling Volkswagens" (they anticipate winning 3.1 percent of the total U.S. sales—again over 200,000 vehicles), it's just that the Karmann-Ghia has reached such a state of refinement that mass production techniques have replaced the former almost-hand-built construction with no loss of quality. You'll see a lot more Ghias simply because more are being made and the car will receive a greater share of VW's $4.5 million 1962 advertising budget.

A further indication of the Karmann-Ghia sales push (sales putsch?) is that prices have been reduced by $135 on the coupe and $200 on the convertible so the present POE prices are $2295 for the coupe and $2495 for the convertible. Prices on cars delivered in Germany have been similarly cut so they now list at $1725 for the coupe and $1900 for the convertible, making the prospect of European delivery even more attractive.

In case you think 200,000 vehicles is a lot for VW's 650 dealers to move, the popularity of the car (in most places they're still being sold at full list price and with a waiting period) has led the firm to seek even more dealers so that in 1962 there will be at least 700 dealers in 50 states (there are over 5400 VW dealers in more than 120 countries). On an average, each of the American dealers sells over 300 units per year—more than any imported car dealer and even greater than Chevrolet's estimated 222 cars per dealer. Of course Chevy's volume far exceeds VW's with 784,103 vehicles registered in the first half of 1961 compared to VW's 87,904. It's possible VW will see greater sales per dealer in 1961 and 1962, except that new dealers are being recruited all the time. Dealer organization and serv-

Road Research Report:
KARMANN-GHIA
VOLKSWAGEN

ice are cornerstones of VW's success in this country. We'll go into them further after we've had a look at the car.

WOLFSBURG IN CHIC CLOTHING

If the Volkswagen is "the poor man's Porsche," then the Ghia could be termed the Carrera for the Common Man. This may be misleading since beneath the fancy wraps beats a heart that's pure Wolfsburg. The engine, chassis and everything else are standard Volkswagen. The only thing different is the body.

Nevertheless, the sleeker shape permits a higher top speed in spite of the increased weight over the standard car. The lower seating position imparts in feel and fact a handiness to the Ghia that the VW doesn't possess.

The Ghia is powered by VW's carefully evolved opposed four-cylinder overhead valve engine. Its current version, introduced late in 1960, develops 40 bhp at 3900 rpm and 61 lb-ft of torque at 2000 rpm. The first post-war VWs had only 985 cc engines, developing 23½ bhp (DIN). The 1131 cc unit was introduced in 1948 and raised the power to 25 bhp (DIN), remaining in production until 1954 when the capacity reached the current specifications with the power increase at the time giving rise to bumper stickers reading "You Have Just Been Passed By 36 Horsepower." The 1960 power increase was accomplished by raising the compression ratio as well as the peak rpm.

Compared to Ghias imported before August, 1960, the latest cars have a sudsier feel: acceleration is brisker feeling with more torque in low ranges and its flexibility is improved. Coupled with the all-synchro transmission introduced at the same time, the car doesn't require rowing motions with the shift lever to propel it, but responds well to an enthusiast's driving techniques.

TECHNICAL NOVELTIES

A key starter introduced on the later cars, and continued for 1962, incorporates a clever lock-out device to prevent engaging the starter when the engine is running. A mechanical arrangement in the switch itself, it takes the form of a ratchet which allows the key to be turned once all the way to the right, energizing the starter, but requiring that the ignition be turned off before the cycle can be repeated. A reason cited for its inclusion is that while raising the power, VW engineers also reduced the noise, claiming that the newer engines are so quiet some owners may not realize they're running. It is quieter than earlier engines, at least when new, but as mileage rolls up the well-known sounds become more apparent—more to bystanders than occupants though, because of extensive sound-deadening.

After cranking a few revolutions, the engine percolates at a fast idle on the automatic choke, the revs tapering as operating temperatures are reached. Warm-up is quite rapid. In chilly weather a peculiar high pitched squeak may be heard but this is usually just the throw-out bearing and the noise disappears when the clutch is operated and the engine warms up.

The VW has one of the few engines we know of (Porsche is another) that will actually cool down (stabilize, not run cold) in heavy traffic. After barreling along a turnpike or through the mountains you may find the oil light flickering, but a few minutes at low revs will cause it to go out and, of course, driving at moderate speeds will do the same. The only precaution is that the fan belt be in good condition and properly tensioned. To eliminate the embarrassment of being stuck in the middle of nowhere when a belt brakes, VW includes a spare in the tool kit. Naturally, most gas stations carry suitable belts anyway and installation is a simple procedure but there is no undue strain on belt so its life should be equal to that of any other car. If the belt should break, the ignition warning light would come on immediately since the belt drives the generator.

ANTI-SMOG DEVICE

Just as overheating is a very unlikely experience, so is freezing. Carburetor icing, something that used to be an occasional problem in cool, slightly humid weather with the older VW engines, is virtually eliminated thanks to a redesigned air filter which, when the engine is cold, takes heated air from the left-hand exhaust manifold. When the engine is warm and being driven at high speed (or in summer) cold air is drawn in to provide maximum power. The air cleaner intake has a flap which controls the source of the intake air. An innovation on the 1962 cars is the addition of a flexible tube from the crankcase breather pipe back to the intake manifold as part of VW's participation in anti-air pollution campaigns. The valve covers have no breathers.

Because of the low piston speed, there is no need to break in the car in the accepted sense of the phrase. It can be driven normally and at high speeds right from the start. The only thing VW cautions against is prolonged foot-on-the-floorboard touring and, more important: no lugging. The latter would impose a heavy load on the engine without providing enough cooling air. The cars are delivered with only 1½ quarts of oil in the pan because the lubricant is for break-in purposes only and is drained at the 300-mile mark. Oil economy is one of the strongest virtues of the engine, not a drop (at least not a drop that was evident from dip stick readings) being used during all our testing.

The number of VWs with over 100,000 miles on the odometer is steadily increasing in the U.S., many of them with no major overhauls having been done. Owners in Germany receive a small badge from Volkswagenwerk when their cars pass the 100,000 kilometer mark, (Continued on page 37)

The jacking points are nearly central, and the jack lifts both front and rear wheels. Jack is stored in front trunk.

Accommodation for cabin baggage is generous, and we found the rear window defroster a very useful standard feature.

The opening rear quarter-windows give good ventilation but raise noise level.

Simplicity is the keynote of the interior in typical VW fashion, but the Ghia displays greater detail refinement.

The grip ring for hubcap removal fits into two holes on the edge of the cap.

Road Research Report: VOLKSWAGEN KARMANN GHIA

Importer: Volkswagen of America, Inc.
Englewood Cliffs, New Jersey

Number of U.S. Dealers: 650

Planned annual production: 10,500 (estimated number of Karmann Ghias for U.S. delivery in 1962)

Value of spare parts in U.S.: $40 million

⅛ SCALE

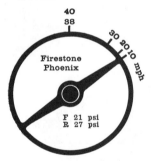

Steering Behavior

Wheel position to maintain 400-foot circle at speeds indicated.

F 21 psi
R 27 psi

Firestone Phoenix

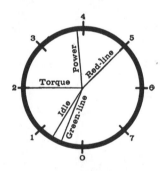

Engine Flexibility
RPM in thousands

PRICES

Basic Price..........................$2295 ($2495, convertible)
Options fitted on test car:
  Whitewall tires .... $ 35
  Undercoating   28
  Sideview mirror   8
Total price of car as tested...............................   2366
Options available:
  Push-button radio   82
  Floor mats   12
  Bumper reinforcements   25
  Heater booster   30
  Cigarette lighter/maplight   7
  Seat belts   34
  Luggage rack   25

OPERATING SCHEDULE

Fuel recommended ....Regular
Mileage ....30-36 mpg
Range on 11-gallon tank........................330-400 miles
OIL recommended...............................SAE 10W-30
Crankcase capacity.............................2.65 quarts
Change at intervals of............................3000 miles
Number of grease fittings 10
Lubrication interval...............................300 miles
Most frequent maintenance..........Check fan belt tension: 300 miles

ENGINE

Displacement..........................72.7 cu in, 1192 cc
Dimensions................4-cyl, 3.03 in bore, 2.52 in stroke
Valve gear: Overhead valves operated by pushrods and rockers from central camshaft below crankshaft
Compression ratio..............................7.0 to one
Power (SAE)............................40 bhp @ 3900 rpm
Torque..............................61 lb-ft @ 2000 rpm
Usable range of engine speeds................500-5000 rpm
Corrected piston speed @ 3900 rpm.................1727 fpm
Fuel recommended ....Regular
Mileage ....30-36 mpg
Range on 11-gallon tank........................330-400 miles

CHASSIS

Wheelbase ....94.5 in
Tread.......................F 51½ in, R 50¾ in
Length ....163 in
Ground clearance.......................................6 in
Suspension: F, ind., trailing links and transverse torsion bars, telescopic shock absorbers, anti-roll bar; R, ind., swing axles, torsion bars, telescopic shock absorbers
Turns, lock to lock.....................................2¾
Turning circle diameter, between curbs...........L 35, R 32 ft.
Tire and rim size..................5.60 x 15, 15 x 4J
Pressures recommended
...............Normal, F 16, R 20 psi; High-speed, F 17, R 23 psi
Brakes, type, swept area....9-in drums, Lockheed hydraulic, 157 sq in
Curb weight (full tank)...........................1790 lbs
Percentage on the driving wheels............................59

DRIVE TRAIN

Gear	Synchro	Ratio	Step	Overall	Mph per 1000 rpm
Rev	No	3.88	——	17.00	—4.2
1st	Yes	3.80	84%	16.63	4.3
2nd	Yes	2.06	56%	9.01	8.0
3rd	Yes	1.32	48%	5.78	12.3
4th	Yes	0.89		3.89	18.6

Final drive ratio: 4.375 to one.

ACCELERATION

Zero to	Seconds
30 mph	8.5
40 mph	13.6
50 mph	19.5
60 mph	27.6
Standing ¼-mile	25.5

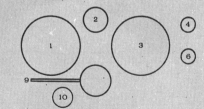

1 Speedometer
2 Fuel gauge
3 Clock
4 Windshield wiper and washer
5 Ignition and starting key
6 Headlight and instrument light switch
7 Fresh air and defroster control — left
8 Fresh air and defroster control — right
9 Turn signal lever and headlight flasher
10 Front trunk lock control

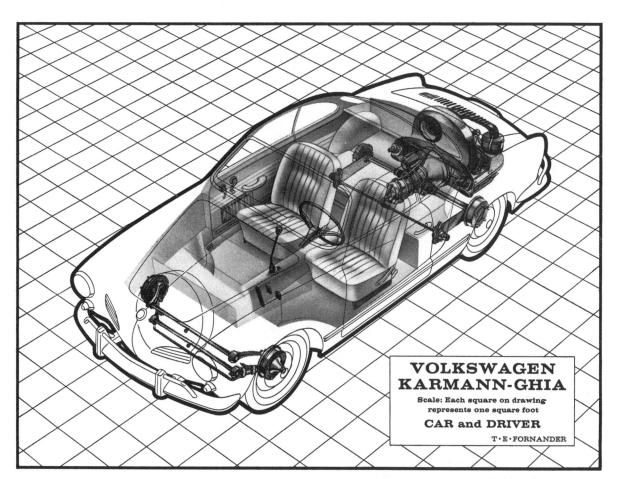

**VOLKSWAGEN
KARMANN-GHIA**

Scale: Each square on drawing
represents one square foot

CAR and DRIVER

T·E·FORNANDER

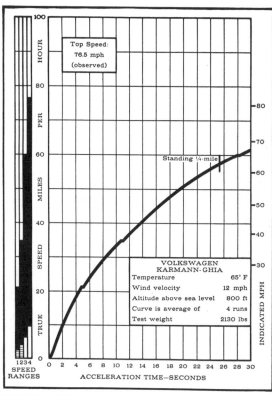

Top Speed:
76.5 mph
(observed)

Standing ¼-mile

VOLKSWAGEN KARMANN-GHIA	
Temperature	65° F
Wind velocity	12 mph
Altitude above sea level	800 ft
Curve is average of	4 runs
Test weight	2130 lbs

HOUR   PER   MILES   SPEED   TRUE

INDICATED MPH

1234
SPEED
RANGES

ACCELERATION TIME—SECONDS

When a naked chassis needs to be clothed motor moguls who care usually say, ". . . HAVE KARMANN MAKE THE BODY."

by ALEX WALORDY

● Rare is the enthusiast who has not contemplated a special car body of his own specifications, preferably a do-it-yourself project. Most of these specials stay in the domain of wishful thinking, some are abandoned at varying stages of completion, while a select few do reach eminent success. Generally, though, the thinking resolves itself into a trip to the local merchant of limited production cars in quest of something exotic, pleasing to the eye, and within reach of the pocketbook. At about this time, you are just bound to encounter one of Herr Karmann's products.

Karmann, in Osnabruck? Oh yes, you must mean Karmann-Ghia? Actually, there is much more to it than that since, in addition to coupe and convertible Karmann-Ghias, they also make bodies for Porsche coupes and Volkswagen convertibles, assembling the completed vehicles. However, the body work represents only a third of the plant's activities. Making body stampings both for their own and outside use and the manufacture of stamping dies account for the other two-thirds.

To transform a flat piece of sheet metal into the intricate shape of a fender or a hood, calls for presses that develop as much as 2,000 *tons* of pressure. The shape itself is provided by a set of dies which support the metal and draw (stretch) it to the right contours. Where the draw is fairly deep, the metal may have to be formed in several successive steps. Sometimes, to avoid wrinkles, an outer ring descends on the sheet metal, gripping it, and then the punch comes down, drawing a final shape.

Since a set of dies for a roof or a floor panel can weigh in the forty ton bracket, but must be finished to high accuracy, you can see that companies with the necessary manufacturing skills and facilities are highly prized by auto makers. Karmann, for instance, supplies dies to most of the European car builders and their list

Molds are taken from the master parts, which are never to be used in production in order to retain their accuracy.

Complete replicas of all parts to be stamped are precision-made in wood or plastic, and are kept for master patterns.

of customers reads like a "Who's Who." Names range from Alfa to Volkswagen, and include, among others, Fiat, Simca, Citroën, Lancia, and Ford in Dearborn, Dagenham, and Cologne.

Sketches or a clay model mark the modest beginnings of a new car in a styling studio. Once the overall design is approved for future production, engineering drawings and a scale model follow. When a company like Karmann receives an order for dies to produce the car, they translate the manufacturer's data into finished drawings of each individual part to be stamped. Their wood working shop then makes up impeccably finished master models of projected stampings. These models are never used in actual production, since they will act as standards for the dies and the finished product. However, plastic or plaster copies are made of the master, as needed. Karmann also makes an overall model of the car to check the fit of individual components, such as doors, hoods, windows, etc. Last but not least, the wood working shop produces mock-ups of the parts to serve as foundry patterns for the cast steel dies.

The rough casting, as it returns from the foundry, must be transformed into an exact replica of the master.

To accomplish this, a plastic mold is first taken of the master. This represents, in effect, a "copy negative." "Copy" and casting are placed on a Keller machine (copying machine). The Keller machine then automatically mills an approximate replica of the pattern into the casting.

The next step is to make a punch which is a matching replica of the die. Here again, we go through the steps with a pattern, a casting, a trip to the Keller machine, and more hand finishing on the spotting press. The die set is, of course, carefully cross checked against the master pattern.

To speed up proceedings, very elaborate copying machines are in use at Karmann's, which can make both a right and a left fender from a pattern of just a left one. Increased accuracy is gained by supporting the pattern casts with an elaborate lattice work of piping. For very small production runs, and for experimental models, the dies are often made of a plastic material such as Kirksite, instead of steel.

Karmann assembles four different types of cars, at the rate of approximately 200 a day, a number that seems tiny when compared to daily figures of one of the "Big Three." This obviously calls for different approaches and generally involves more individual skills on the part of the assemblers. For instance, it would not pay to make up special dies for the rear quarter panel of a Volks convertible, so they cut off the window section of a VW sedan quarter panel, and tack on a small section to fit the belt line of the contour on the convertible. It's hand welded, dinged out, and looks as good as any one piece stamping. Junctions that would normally be covered by moldings are carefully hand hammered into position, and leaded on the line.

This unusual amount of individual skill calls for a rather elaborate training program. Karmann believes in starting them young, and operates a vocational school right at the factory for youngsters from 14 to 18, teaching them the fine points of tool making and experimental sheet metal work in addition to their normal high school curriculum. Some of the classroom projects we saw carried out by students would make an experienced tool maker blink.

It's in shops like Karmann's that the phrase "old-world craftsmanship" is kept alive. And it's nice to remember, in these days of rampant automation, that at the beginning of a new car there's a man with an idea and the skilled hands to implement it.

The die is hand finished in a spotting press and comes out a mirror replica of the punch. The grooves are tool marks.

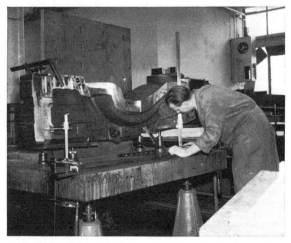

A finished die is carefully checked against the master. Whenever a die is worn, it is welded up and resurfaced.

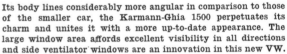

Its body lines considerably more angular in comparison to those of the smaller car, the Karmann-Ghia 1500 perpetuates its charm and unites it with a more up-to-date appearance. The large window area affords excellent visibility in all directions and side ventilator windows are an innovation in this new VW.

KARMANN CLOTHES THE VW-1500

● For all those who want the virtues of the standard Volkswagen encased in a better-looking body, the Karmann-Ghia, well-known in both convertible and coupe versions, has been the answer. It seemed only a question of time, therefore, when the VW-1500 was announced last April, before Karmann would produce a body for it. It was first shown at the Frankfurt Auto Show last September, both a coupe and a convertible being on the stand. In addition, Wolfsburg chose the same occasion to introduce its own four-seater convertible.

The 91-cube engine develops 53 bhp at 4000 rpm, with gear ratios unchanged from the Karmann-Ghia in this issue's Road Research Report, which means that acceleration is improved rather than maximum speed.

In addition to providing more pleasing lines than the now-outdated beetle, the VW 1500 body shape gives more baggage room in the back as well as in the trunk under the front hood.

The space utilization in the Karmann-Ghia 1500 may not be quite up to Wolfsburg ideals but the comfort for two front seat passengers is decidedly superior. It is still not known when the 1500 will be introduced in the American market.

ROAD RESEARCH REPORT: KARMANN-GHIA VOLKSWAGEN

▶30

stating that they have done so. Perhaps with good reason, the badge consists of a VW emblem and a St. Christopher medal.

The clutch pressures are quite low, almost having a soft feel, yet engagement is smooth, if a trifle low. Popping it out for fast acceleration causes a noticeable sagging feel and a little slippage before the car gets rolling. Fast upshifts at high speeds give the same sensation, but the slippage is more apparent than real, for it bites solidly on downshifts, even the brutal kind without feathering the gas pedal. Its action is quick; double-clutching (not necessary, but a little easier on the mechanism) can be done as quickly as you can move your feet; just watch out for the nearby brake pedal.

The gearbox itself has ratios well suited to the car, although first could usefully be a little higher. The pattern is a near-H and the linkage is smooth and crisp. It has good feel, but the speed with which shifts can be executed seems to vary from car to car and our test vehicle wasn't one of the fastest shifting we've used. A noticeable amount of noise is carried from the transmission up the spine of the car and some of the noises are quite peculiar, but perfectly natural. The convenience of the lever is a matter of preference. With the seat fully rearward, the lever could well have been located two or three inches farther back.

WINDTUNNEL EXPERIENCE

The Ghia's acceleration is far from fierce, but it's more than just adequate. Passing still requires a little planning and climbing a particularly long, steep hill sees the speedometer needle sink steadily. The cruising speed is whatever you and the law decide on up to the 76 mph top speed of the car. Touring at these speeds, even allowing for temporary baulking by traffic and steep hills, is easy on the nerves and the car. Part of this is due to the almost total lack of wind noise. On our demonstrator, the side windows didn't seat perfectly on their rubber mouldings so there was some whistle from the one on the passenger's side. Ventilation is excellent, although vent-panes in the side windows would be useful. Two vents, working through the defrosters, are individually controllable and admit vast quantities of fresh air with no draft. The hinged rear quarter windows give an extractor effect and even with the side windows down, breezes within the cockpit take the form of gentle zephyrs. Very relaxing. However, the Ghia is designed mainly with two people in mind so the unfortunates who get stuck in the emergency seat, find it's like sitting in the wash of a windtunnel when the windows are all open.

The brakes on our car suffered from a low pedal. This is not typical, but could indicate that VOA public relations department doesn't practice what it preaches to the half-million VW owners in this country: Get your car serviced regularly from an authorized dealer. Nevertheless, based on our experience with other Ghias, pedal pressures are low and stopping distance is a function of how hard you press on the pedal. The four wheels can be locked up tight with

a minimum of nose dipping and no loss of control. Even with the brakes on our car, panic stops were made in perfect safety.

SIDEWINDS AND HANDLING

There was some play at neutral lock on our test car's steering, requiring fairly large wheel movements to counteract any tail wagging at turnpike speeds in heavy crosswinds. However, once the play was taken up, the steering was very light and fairly direct. Its fast enough for a good driver to enjoy, yet not so skittish that an inexperienced one will steer too hard and find the rear end passing him on a curve. Return action is moderate and there is good feel even up to the limit of adhesion. Virtually no road reaction is transmitted to the driver. One of the delights of any Volkswagen is that it feels right at home at 40 or 50 mph on rutted dirt roads. A steering damper built into the linkage reduces road shocks drastically.

As we hinted, tracking on the straight is affected by sidewinds. This is to be expected with 59% of the weight on the back wheels and no significant built-in understeer. If anything, the Ghia is nearly neutral until, as the limit is approached, it changes to oversteer. The car responds well to steering wheel movements, but unless the gearbox is used or the speeds are only moderate, the ability to use the throttle for cornering is not great mainly due to the lack of brute power. The rear end behavior is predictable—once you have experienced its action a few times at the breakaway point it's a useful phenomenon. Novices are cautioned, however, to tread lightly until they get the hang of things, but it's only fair to add that if you do get in trouble you can only blame yourself, since the car is with you all the way.

Our car used Firestone Phoenix tires which even with very high pressures squealed lustily on corners—and not even rapidly negotiated corners at that. No lean is apparent to the driver, although the car is suspended in such a way that big bumps are reduced in intensity and the front anti-roll bar helps keep the inside rear wheel from lifting. The ride is firm, but not uncomfortable. It's resilient but not mushy. Damping of all bumps is excellent although there is some pitching on undulating surfaces. The isolation of road noise through undercoating and separation of suspension components from the main chassis members is similarly very good. The car has an extremely solid feel—it's one of the few cars we've tested that had absolutely no rattles or squeaks despite the treatment it received.

The doors open wide, but the low roof line naturally impedes access which requires a knack to be accomplished easily, particularly with the seat well forward. A two-position holder keeps the doors open. The individual seats are firm, not at all bucket-like yet are untiring to sit in for long periods. They are adjustable for fore and aft movement and the seat back is adjustable for three angles of rake.

OPTIMUM OPTICS

Because of the intrusion of the front wheel wells into the passenger compartment, the driving position is almost side-saddle, with the driver's legs pointing toward the middle of the car. The wheel was a little too close for most long-armed drivers, but arm and shoulder room was very good. Visibility forward is good, improved by the narrow door posts. To the rear, it is similarly panoramic, but

the outside rear view mirror is needed. Our car was a blue-green color and the rear "parcel shelf" reflected on the back window, reducing visibility in some lights. Similarly, the steering column was visible as a ghost before the driver's eyes at times, including at night when the panel lights were on bright.

The steering wheel itself is a good size and has a dished hub. The pedals and the space between them is on the skimpy side, although the relative location of the brake and organ accelerator pedals is perfect for heel and toeing. The ratchet hand brake lever is located between the seats, having a pushbutton which must be held to release it.

Of instruments, there are three. All are round and very legible. The big speedometer occupies the left hand position. It has on it the ignition warning light, the low oil pressure light and the high beam indicator. Equipped with an odometer, it has no tenths of miles column or resettable trip odometer, but one is available from VDO which is identical to stock with the addition of this feature. There is an apocryphal tale to the effect that the tenths of miles column is actually built into the stock VW instrument, but that it is shielded so that all that need be done is to clip away part of the face. We haven't checked this out. The fuel gauge is another VDO unit, a mechanical one. It shows how much gas is in the tank—what more could you want from a fuel gauge? Since its inclusion the reserve tap has been eliminated from the toeboard. The third dial, just as big as the speedometer, is the electric clock. Wouldn't it have been more in keeping with the sporting character of the Ghia if that space had been used to house a tachometer? However, a mechanical tachometer, though not as big as the clock, is available from VDO. The instruments are located in recessed holes to keep their lights from straying to the windshield. This works, but as we said the light hitting the steering column does become annoying and requires the rheostat switch to be turned down.

The windshield washer is a new design. It features a water tank with a tire valve on it. After the tank is filled with water to the recommended level, 35 psi of air is pumped in. A button, located in the center of the wiper switch, is pressed to squirt the water and it will keep squirting as long as the button is pressed until the water runs dry or the air pressure becomes too low.

Almost everything on the VW works with what the old cliché terms Teutonic thoroughness. The wiper and light switches sort of glide on and off with a smooth precision. It's the same with the two vents on the dashboard, the hood and engine lid releases and the handbrake. Almost everything, we said. The heater control is smooth, yes, precise, yes, but takes about 15 hand movements to rotate it the number of turns required to set it on full blast or turn it off. The side windows too require too many turns and one of the rear vent windows was very stiff to use. The blanking flaps on the heater outlets are not very positive either, sometimes sticking in operation.

What the VW ads say about air tightness is no exaggeration. It is easier to close a door when a window is slightly open. The car was absolutely free from leaks and the trunk and engine compartment too are well sealed. The heater seems to have been improved so that even at relatively slow speeds heat is adequate. We didn't have an opportunity to drive the test car in below zero

CONTINUED ON PAGE 65

KARMANN GHIA VW

A sleeker wagen für fewer (und richer) volks

Ⓥ CARS MAY COME AND CARS MAY GO, but the Volkswagen goes on forever—or so it seems. And the rich cousin to the lowly people's car, the Ghia-designed, Karmann-built coupe and convertible seems also to go on forever.

Road & Track conducted its first road test of a Karmann-Ghia Volkswagen in April 1956, shortly after its introduction, and said "This coupe is actually available for purchase, providing the customer is well supplied with patience." The statement is still valid today, although the wait is not quite as long as it was in 1956.

We, too, have been waiting, not to buy one, but to test a newer model, never expecting that, like the sedan, it would be continued (seemingly forever) with little or no change. There is, of course, a Karmann version of the VW-1500, which we will get to later, but this test car is merely a refined version of the one tested six years ago.

Since the Karmann-Ghia's introduction, it has almost seemed an orphan, due to the lack of effort put into advertising or promoting it, even though sold and serviced by all VW agencies. It is, in essence, a standard Volkswagen with more attractive bodywork (for approximately $700 dollars extra).

Few changes have been made other than the usual and expected refinements that were also added to the sedan. In fact, every change and/or improvement made to the sedan has been incorporated into the K-G (wherever applicable) at the same time. Most important was the change in 1960 from 36 to 40 bhp, a seemingly insignificant amount when viewed as only a four horsepower gain, but it represents an 11% increase, which made the difference between barely adequate performance and that which is very satisfactory.

The addition of a fuel gauge in the 1962 models was a move that was long overdue in the minds of many Volkswagen owners, and yet we've talked to some who still prefer the old tried and true reserve tank with manual control switch. And then there are those who would like both—but some people are never satisfied.

The seating in the K-G is considerably more comfortable and sporting than it is in the sedan, but entrance and exit are more difficult due to the low build of the car. Headroom is also more limited because of the body styling, but this is the price one pays for the better appearance; a price apparently thought to be well worthwhile by more than 9000 K-G buyers (new cars only) in 1961. There is no shortage of foot room because of the great amount of seat adjustment. The front wheel wells encroach into the passenger area, causing the pedals to be mounted slightly offset to the right. This puts the new driver off a bit, but he soon gets used to it.

Other than the obviously different seating position, the Karmann-Ghia drives and handles like its production counterpart, just as you'd expect. All the familiar Volkswagen noises, peculiarities and attributes are there, and in about the same amount.

For some unexplainable reason, one *feels* as though the coupe is performing much better than the sedan. In top speed it does—due to the lower frontal area and better shape. Acceleration, though, is slightly better on the sedan because of a weight penalty of about 70 lb for the coupe. The cross-over point on the two cars' acceleration curves is at very nearly the top speed of the sedan. As with any make of car, exceptions can always be found and there will occasionally be a Volkswagen sedan that will run rings around a coupe, but it isn't normal.

The lower frontal area and subsequent higher top speed pay off in another way too—the ability to maintain a higher cruising speed. Because of the engine design and factory selected gear ratios the cruising speed of a Volkswagen is the same as its top speed. This, then, automatically gives the coupe about 8 mph advantage over the sedan for sustained and reliable cruising.

We've heard many times about how the K-G coupe will be less disturbed (than the sedan) by cross winds because of its lower silhouette. We're not convinced. We drove the coupe in a hard, gusty off-shore wind along the California coast and couldn't detect any great difference between the two, if there was any.

It would seem that the coupe should be less affected in a wind because of the lower profile (and less mass) presented to the wind, but a car's handling characteristics in a cross wind are determined, to a great extent, by the relation of the center of gravity to the center of pressure: i.e. if the c.g. is forward of the c.p., then the car will be [theoretically] stable. If the c.g. is aft of the c.p. then the car will be [relatively] unstable. The rearward weight bias of the coupe (58.3%) is greater than the sedan's 57% and this, coupled with the coupe's c.p. being farther forward, reduces the advantage of the smaller total area.

The adverse effect of cross winds (which incidentally

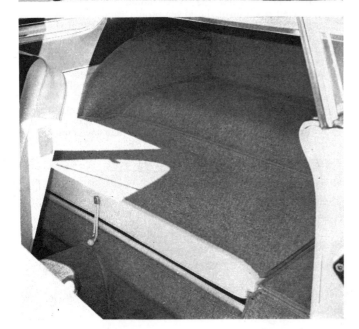

apply to most tail-heavy cars) is the only strong criticism we have of the Karmann-Ghia coupe's handling. In any other situation the car is very pleasant to drive.

In expressway driving there is very little to choose between one car and another—they all seem good—but when the going gets rough, cars with independent 4-wheel suspension show a marked superiority over their competitors. The K-G is no exception and takes freeways, ruts, bumps and potholes with equal aplomb and little pitching as long as the dampers are good. The pitching action that does exist is almost identical to that of the parent Volkswagen but is not as noticeable because the driver sits closer to the pitch axis than he would in the sedan, where pitching movement is amplified.

In designing the Volkswagen, Dr. Ing. Ferdinand Porsche utilized a front and rear suspension arrangement not too much different from that of the famous Auto-Union racing cars which came from his drawing boards and which also were powered by rear-mounted engines.

All current Volkswagens retain the swing axle rear end (some sort of independent suspension is necessary with a rear-mounted engine) and parallel trailing arm front suspension that has been a design feature of the Volkswagen since the first model. Transverse torsion bar springs are used all around, laminated square in front, round at the rear, with tubular hydraulic shock dampers.

As with all cars carrying a majority of the total vehicle weight on the rear wheels, there is a tendency to oversteer —the degree of oversteer being determined to a great extent by the type of suspension employed.

To help counteract the oversteering characteristics, a problem so serious on the racing cars that few drivers could ever make full use of the brutish 600 bhp available, Dr. Porsche designed a front suspension with the roll center at ground level, which contributes to an understeer condition. This minimizes, but does not completely negate, the oversteer caused by the rear end weight bias.

Casting all theory and speculation aside, the Karmann Ghia is still a Volkswagen, with all that the phrase implies. It will carry two people, and their luggage, in reasonable comfort at a reasonable speed, and with economy and reliability. Not much more could be asked of it.

ROAD TEST
VW KARMANN GHIA

SCALE: 10" DIVISIONS

DIMENSIONS

Wheelbase, in	94.5
Tread, f and r	51.4/50.7
Over-all length, in	163
width	64.2
height	59.2
equivalent vol, cu ft	359
Frontal area, sq ft	21.2
Ground clearance, in	6.0
Steering ratio, o/a	14.2
turns, lock to lock	2.4
turning circle, ft	37
Hip room, front	2 x 22.5
Hip room, rear	41.5
Pedal to seat back, max	41.0
Floor to ground	9.7

CALCULATED DATA

Lb/hp (test wt)	52.7
Cu ft/ton mile	63.7
Mph/1000 rpm (4th)	18.8
Engine revs/mile	3190
Piston travel, ft/mile	1340
Rpm @ 2500 ft/min	5950
equivalent mph	112
R&T wear index	42.8

SPECIFICATIONS

List price	$2295
Curb weight, lb	1750
Test weight	2110
distribution, %	41.7/58.3
Tire size	5.60-15
Brake swept area	156
Engine type	flat 4, ohv
Bore & stroke	3.03 x 2.52
Displacement, cc	1192
cu in	72.7
Compression ratio	7.0
Bhp @ rpm	40 @ 3900
equivalent mph	73.4
Torque, lb-ft	64 @ 2400
equivalent mph	45.1

GEAR RATIOS

4th (0.89)	3.89
3rd (1.32)	5.77
2nd (2.06)	9.01
1st (3.80)	16.6

SPEEDOMETER ERROR

30 mph	actual, 28.8
60 mph	56.0

PERFORMANCE

Top speed (4th), mph	75
best timed run	75.8
3rd (4500)	57
2nd (4550)	37
1st (4550)	20

FUEL CONSUMPTION

Normal range, mpg	30/35

ACCELERATION

0-30 mph, sec	6.4
0-40	11.4
0-50	17.3
0-60	30.0
0-70	54.5
0-80	
0-100	
Standing ¼ mile	22.7
speed at end	55

TAPLEY DATA

4th, lb/ton @ mph	140 @ 38
3rd	220 @ 32
2nd	350 @ 26
Total drag at 60 mph, lb	105

ENGINE SPEED IN GEARS

4th
3rd
2nd
1st

2000 3000 4000 5000
ENGINE SPEED IN RPM

ACCELERATION & COASTING

90
80
70
60
50
40
30
20
10

SS¼
4th
3rd
2nd
1st

MPH   5 10 15 20 25 30 35 40 45
ELAPSED TIME IN SECONDS

Tested in Germany

Our Continental correspondent tries the most handsome VW so far, and finds it the most desirable also. Good performance and handling, but restricted luggage capacity are amongst his findings.

BOXER WITH A PEDIGREE

VW 1500 KARMANN-GHIA

The prevalent theory that one Volkswagen is just like the next, and both undistinguishable from the first one, simply won't stand up in face of the VW 1500 Karmann-Ghia. This is a different Volkswagen—particularly compared to the 1.5-litre saloon—with roots in the original Karmann-Ghia. But it is not the same as that one either. Not only is it different, I found it more pleasant in nearly every respect.

To my taste they have retained the plus points like detail finish, durability and cachet, adding style, at least the illusion of power, comfort and just a taste of individuality which was lacking in previous cars. When testing the 1.5 saloon some months back I still missed this dash of paprika, but the deep red 1500 KG had it, from the first moment. You can't explain such things but the subjective impression won't be excised.

For the record, the larger Karmann-Ghia—the 1200 KG remains very much in evidence around Wolfsburg—bears the same relationship to the 1500 saloon as the smaller two-seater did to the "beetle" when new. All mechanical parts are identical. Volkswagen even denies vehemently the persistent rumour that they side-track particularly good engines on the production line and put them in the Ghia models.

I won't get into that argument—suffice it to say that the test car we collected from Wolfsburg had one of the hottest "stock" VW engines it's been my pleasure to use. It didn't feel oversize or anything like that, just perfectly balanced, perfectly run-in and exceptionally willing. This could be part of our pleasure. It had nearly 12,000 miles on the clock, incidentally, which is about right for a "Volks". They come into their own by then.

The mileage, all of it spent in hard testing, underlined the quality control at Osnabrueck where the Karmann factory makes these shells. While I questioned the perfectionist myth in the 1500, Saloon, this KG lived up to all Volkswagen boasts. Doors closed like new, seats slid at a finger touch, lids opened and closed first try and the instruments all did honest duty. Only the gearbox, or rather

LEFT. *The three-quarter rear view of the KG 1500 is one of its most handsome aspects. Ornamentation is restrained and dignified, and visibility through the large glass area quite exceptional.*

TOP PICTURE. *When rushed through fast curves the VW 1500 Karmann-Ghia holds its line well with little roll, and without excessive oversteer.*

ABOVE. *Seats and instruments are the right shape, and the gear-lever is well positioned. The driving position would be even better without a dished steering wheel. Central handbrake is robust, and there are useful door pockets.*

The "extra" lights are not greatly effective in an otherwise well-designed car. The small ones are too high to be efficient in fog, and too far from the main headlights to be part of a true "quad" arrangement. They are useful for town driving.

SPECIFICATION

PERFORMANCE

0–20 m.p.h.	3.0 sec.	0–50 m.p.h.	13.2 sec.
0–30 m.p.h.	5.7 sec.	0–60 m.p.h.	19.7 sec.
0–40 m.p.h.	9.2 sec.	0–70 m.p.h.	28.2 sec.

Speedometer error: 3.9 per cent.
Maximum speeds: First, 30 m.p.h.; Second, 48 m.p.h.; Third, 67 m.p.h.; Top, 78.9 m.p.h.
Test mileage: 743 miles.

ENGINE
VW four-cylinder, air-cooled, horizontally-opposed, overhead valves (pushrod) and central camshaft. Bore: 83 mm. Stroke: 69 mm. Cubic capacity: 1,493 c.c. Compression ratio: 7.2 : 1. Power output: 53 (SAE) b.h.p. at 4,000 r.p.m. Maximum torque: 83.2 lb. ft. at 2,000 r.p.m. Single Solex 32 PHN horizontal carburetter. 6-volt lighting and starting.

TRANSMISSION
Four-speed gearbox with synchromesh on all forward ratios. Lever on floor tunnel. Gearbox ratios: First, 3.80; Second, 2.06; Third, 1.32; Top, 0.89; Reverse, 3.88. Final-drive ratio: 4.125 : 1.

SUSPENSION
Independent front with twin trailing arms and torsion bars, telescopic dampers. Independent rear with single trailing arms and torsion bars, telescopic dampers.

BRAKES
Drum brakes front and rear with 128.7 sq. in. lining area. Steel, bolt-on wheels.

BRAKING FIGURES
Using Bowmonk Dynometer. From 30 m.p.h., 92 per cent = 34.2 feet.

DIMENSIONS
Wheelbase: 7 ft. 9.5 in. Track: front, 4 ft. 3.6 in.; rear, 4 ft. 5.2 in. Length: 14 ft. 0.5 in. Width: 5 ft. 3.8 in. Height: 4 ft. 5.4 in. Turning circle: 35 ft. Ground clearance: 6 in. Dry weight: 17 cwt. Fuel capacity: 8.8 gallons (Imp.). Average fuel consumption (autobahn and city): 27.9 m.p.g. Tyres: Firestone Phoenix. 6.00 by 15, tubeless.

WATER PRESSURE TEST
No leakage at any point.

PRICE IN U.K.
To special order and with left-hand drive only. £1,281 7s. 11d. including Purchase Tax. Basic price £1,060.

Front boot. The camera case gives some idea of the limited luggage accommodation. Spare wheel and fuel tank are partitioned from the luggage, in the nose.

Rear boot. The same camera case gives some idea of the equally restricted space at the rear of the car. Although this compartment is immediately over the air-cooled engine, it is very efficiently insulated.

its shift linkages, seemed a touch the worse for hard useage. The famous VW synchromesh was fully on duty but there was a bit of play on downshifts from III to II and just a trace of vagueness between the II and IV sides of the lower gate. If I hadn't expected perfection these points might well have passed unnoticed.

Getting back to the power plant, the test car naturally had the VW disregard for revs., thanks in part to the "overdrive" top gear. But that wasn't the entire story with this particular vehicle. It would wind up apparently forever in the indirects and had a wonderful surge of torque available for *autobahn* passing as well, in a very un-VW manner. The engine actually felt smoother at the top of its power peak than around 2,000 r.p.m.

The drawback to this feeling of tuned power was a permanent tendency to run raggedly at idling speeds after a session of thrashing

Rear compartment is not very comfortable for passengers, but when the back-flap is folded forward the extra space for luggage is practical.

and a positive reluctance to re-start when really warmed up. This too gave the car a personality. It was not a paragon but an eager tool of the driver with quirks and a stout heart.

Two vital areas go with this sort of potential because a car willing to travel flat-out is likely to be driven just that way. If it won't stop or handle you are in trouble. Fortunately the 1500 Karmann-Ghia did both very nicely—particularly stop. I can vouch for brakes that would haul the car to a dead stop from over 70 m.p.h. only a breath short of locking the wheels and do it dead true, with a margin of driver control. It isn't a recommended braking test, but one of the criminally careless German trucks tried to shove us off the *autobahn* and we managed to halt, just two inches from his oblivious tail. Furthermore the brakes recovered far sooner than I did, displayed no pedal loss, and seemed ready to try it again.

The roadholding to go with this sort of game is there too, to a measure they should endeavour to incorporate in the 1.5 saloon. Whereas that model was downright frightening in a cross-wind the 1500 Karmann displayed noticeable sensitivity to gusts but was not tricky. I would rate it the best of the VW line on this score, and one of the two or three best rear-engined cars I know. It isn't as totally unaware of side winds as say a front-drive car but they aren't a problem.

Hustling around the mountains poses no particular problem either, provided you prefer a modicum of oversteer to its opposite. The tail of a KG 1500 certainly goes first if you provoke things sternly but it doesn't whip around, and it really takes an effort to

lose this machine. Granted, you aren't moving at racing speeds on 53 h.p. in nearly a ton of car, but you can drive quickly from one point to another in safety. Roll is negligible.

The luggage load makes very little difference to either side-wind behaviour or cornering for the simple reason that you can't carry enough front or back to make much relative difference. I suppose if you habitually smuggle large amounts of gold brick it would alter the 1500 Ghia's handling, but I seriously doubt if you could get enough travel baggage into either bin to change the weight distribution appreciably.

Both the front and rear compartments are neatly lined, carefully protected from weather —and too shallow to take anything but a regulation suitcase lying flat. Your kit bag is too high for the rear and probably for the front as well. One bonus, incidentally, is that you can carry anything in the tail without heat worries. It was not possible to measure a temperature rise after hard driving. Furthermore, like most two-seaters, the KG has its main luggage provision behind the seats. It isn't protected from casual eyes but the excessively upright back of the rear occasional seat folds down to make an excellent baggage space, with a proper front lip to keep small items from roaming.

Rear passenger space is practically non-existent. With the driver's seat back for proper pedal control there is the minimum legroom, to say nothing of the low ceiling and meagre padding. The two front seat occupants are far better served with excellent padding, good lateral support for hips and shoulders, adjustable backs and well-designed thigh rolls to lessen fatigue. If you could

just get a little further from the steering wheel the car would rank among the best road tourers of them all in driving position. Once ensconced in these seats the driver finds himself facing three round dials. He shouldn't congratulate himself too soon. True, there is a highly accurate and very readable speedometer, and a fuel gauge (in a VW!) but that is about the extent of things. Even if we excuse them for eliminating the water temperature dial on obvious grounds, a little more information beyond the all-or-nothing lights would help. The clock is very pleasant but I'd rather see that space occupied by information I can't get just as easily from my wrist.

Controls like wipers, headlights and sidelights are arranged together flanking two rheostats. One controls dash lighting, a very proper touch by designers who also drive. The other—and it would help at first if they weren't such close friends—gives you infinite wiper speeds. This too is the sort of "extra" that should be fitted to every car but never is. Full marks to Volkswagen for putting their excess profits into details that may not attract impulse buyers but will certainly please their long-term users for the life of the car.

The cockpit of the Volkswagen 1500 KG is arranged for drivers in general. You can reach practically every item but the glove box with a safety belt attached, including the ashtray and cigar lighter. Small things, but proof of the manufacturer's care. Visibility is first-rate in all directions, but that glass expanse can be a bit much on sunny days. You feel a little like a pampered flower in a hothouse, exposed to the sun whether you like it or not.

The 1.5-litre Karmann-Ghia is not a sports car—let's be clear on that point. Despite the beautiful line (I except the voluptuous front swoops), it is not a racer, even by road standards. What's more important is the fact that it gives the driver a feeling he is rapid without getting him in trouble. Unfortunately, on German grounds at least, it does so at a rather inflated price. At approximately ten per cent less money it would be right at the top of your Christmas list to that rich uncle. Even at the current rate—and ignoring purchase taxes and the like—the VW 1500 KG is desirable in a way I haven't found before amongst the flat-four clan.

KARMANN-GHIA VW 1500

A well justified compromise between luxury and economy in a touring car

STORY AND PHOTOS BY HANSJÖRG BENDEL

WHEN THE prototype Karmann-Ghia, built on the chassis of the 1200-cc "beetle," was prepared for the Frankfurt Motor Show in 1955, not many people would have accepted bets on the commercial success of this new venture. After all, the basic VW has not got such an abundance of power that it just asks for a stylish coupe body, and Karmann never made any attempt at tuning its K-G into a "less expensive Porsche." Also, it was immediately apparent that the attractive selling price of the sedan would have to be exceeded substantially, due to smaller production volume as well as because of refined finish. But the customers queued up; just as in the case of the standard model, the K-G confounded its critics by selling by the thousands, and continues to be made, very little changed, at the rate of about 60 per day.

In view of this success, it is hardly surprising that when the VW-1500 became a reality, Wolfsburg and the Karmann works teamed up once more. When the 1500 sedan was unveiled at Frankfurt in the autumn of 1961, new Karmann-Ghia models were ready as well, and in the spring of 1962 the coupe assembly lines were in full operation. Very soon, the convertible sedan 1500 (already shown as a prototype in Frankfurt) will be in production.

Following the practice established with the well-known forerunner, the little-changed platform type chassis of the sedan is combined with a special two door coupe body, *Haute Couture* by Ghia of Turin, with lower build, improved aero-

dynamics and, of course, reduced room for occupants and luggage. On the roads of Europe, K-G 1500 coupes are already familiar, so once again buyers are prepared to pay extra money for a smaller car, just because it is different.

What are they getting for their marks, francs or lire? The first point to discuss might be the question of beauty. Some people like the newcomer very much, whereas others have not a single kind word for it—the rear end is accepted by most, but the front end with its distinctive cat's whiskers framing a narrow-gauge pair of foglamps has been the object of much controversy. One thing seems certain. Beautiful or not, giving the car the "different look" was a compulsory design target which—not many will doubt this—was accomplished.

Because running gear and engine are identical on K-G and sedan, no sensational differences can be expected when driving the coupe, and we rediscovered most of the characteristics already commented upon in the test report of the standard car (R&T, May 1962), which therefore need no recapitulation. The later models which left the assembly lines since August 1962 benefit from improved manufacturing methods and incorporate some detail improvements. On the mechanical side, there are larger intake valves, improved cooling air ducting for the engine and wider brake shoes on the chassis. The body, too, has been generally cleaned up. We drove the original version first, then checked our findings on a late model, which eliminated some points of criticism.

One of the first impressions one gains is that of very nice handling. Just as its more mundane sister, the K-G is fun to drive, highly maneuverable, and always manages to convey an impression of agility and speed superior to the cold figures of the stop watch . . . in other words, a decidedly sporting character and a constant invitation to enjoy one's daily mileage. Broadly speaking, road holding is the same as that of the sedan, with the notable exception of much reduced sensitivity to side winds. Whatever the reason, this is an improvement which is as welcome as it is necessary. Performance is slightly better, particularly at the upper end of the speed range, and certainly satisfactory for the kind of use to be expected from the average owner. The car's effortless running on highways is particularly pleasant and remains an example which several other, more powerful, cars might do well to follow. After descending an Alpine pass in our first test car (which had the original brakes) there was an unmistakable hot smell and brake squeal; but fading was slight, and normal performance and perfect balance were restored quickly. Bearing in mind that this model is definitely not planned for rallies, the larger brake lining surface since introduced should prove ample.

Before setting out now to discuss controls and interior furnishings (where the K-G is of course markedly different from the sedan) we have to qualify: The K-G costs substantially more than the standard sedan and must be looked at as a luxury version. Therefore, it undoubtedly has to be judged by a more exacting yardstick. This explains why we have to lay our finger on some shortcomings which might be perfectly tolerable on the cheaper sedan (some of them are peculiar to the K-G, though) but have no place on a model of higher ambitions.

The first look indoors always goes to the instruments. For

styling reasons, the speedometer has been made smaller than its counterpart on the sedan, and has not been completed by the very useful trip mileage counter familiar on many less expensive cars. For legibility at speed, a full-size instrument would be preferable. The front seats are comfortable and give good support; regrettably, the stylish roof line has dictated such a low seat level that the feel of complete mastery over the car and of perfect visibility is lost. Legroom is generous for all but the most pronounced stretched-arms drivers. On the first cars tried, the rake of the seat backs could not be adjusted with adequate finesse—on the second car we were pleased to find accessible hand wheels in front of the seats, which would have been absolutely perfect if they had required a little less force to turn them.

Seat back tilt is locked during driving, the release catch being totally inaccessible to rear seat passengers. The latter are not well catered to anyway. They are provided with a short bench with a thin seat cushion, entry is difficult and room barely sufficient for children. On top of this, hot air ducts run underneath the rear seat cushion without proper insulation. On both test cars, we found, when driving in hot climate, that this represents an uncontrollable, most unpleasant central heating. When only two are using the car, the rear seat back can be folded flat on the seat, whereupon there is very useful additional luggage room. This situation is similar to the well known space problem on the Porsche coupe, and it is fair to say that for real long-distance traveling, the K-G is a very elegant, practical two seater—not more.

Apart from the suppression of a central cold air inlet (why?), "official" heating arrangements are identical to standard VW practice. Surprisingly, even this luxurious coupe still has no blower to force cold air into the car; VW relies on intake grilles ahead of the windscreen, which become effective only at elevated speeds. Defrosting with the engine switched off is impossible; when starting up from cold, it takes quite a long time before proper visibility is guaranteed. On the latest

KARMANN-GHIA VW 1500

Porsches, these problems can be partly overcome by the installation of a special gasoline-type heater which is an expensive and space-wasting solution. Corvair experience, however, seems to confirm that this difficulty is not easy to overcome with an air-cooled engine living at the rear of the passenger compartment.

There are two sun visors, nicely padded, which conform to the concave shape of the roof. Folded down for use, they still permit a gap of half an inch between roofline and upper edge of sunblinds, so that a low sun spotlights straight into the driver's eyes. Karmann development engineers must have worn very dark sunglasses never to have noticed this fault.

Noise insulation between engine and passenger compartment is improved and, in contrast to some reports we have seen, we have found the wind noise tolerable—though we don't deny that at higher speeds, a lower general noise level would make conversation easier still.

Finally, a few words on details, which are of some importance on a "personalized" car like this. The armrest on the driver's door is in the way when working the steering wheel in a succession of corners; control of the excellent gearbox (on both cars) is less perfect than on the sedan, with a feel of more flexibility in the linkage; the windshield wiper is very

good, fast enough for heavy rain, but noisy on a drying windshield. The front wheel arches intrude too much into the front compartment; this is permissible in the sedan, and seasoned VW drivers are accustomed to brace their left foot against it, but we still feel that on a car of higher price level, this design feature comes in for criticism. Heel-and-toe gearchange is not easy due to the pedal arrangement, though this is admittedly not essential, due to the good synchromesh; ground clearance is generous even for rough going; initial pick-up from slow speeds is even more jerky than on the sedan. Obviously, the ultra low carburetor-cum-manifold arrangement enforced by the rear luggage compartment has posed carburetion and linkage problems which have not been entirely overcome. The rear-view mirror has no anti-dazzle position; the panic bar for the front passenger, very elegantly shaped and practical on the original product, has been changed for the worse.

We are pleased to record that test car number 2, undoubtedly typical of what customers are getting, displayed a high quality level. Its outside finish was perfect, there were no rattles, and the general feel of bank-safe rigidity was what we have come to expect from Karmann as well as from VW. Final criticism: the color combinations of the plastic upholstery seemed garish and tasteless. As the actual material for seats, door trimmings and roof lining appears to be of good quality, carefully fitted, some extra effort in color composition would be well applied.

Summing up: for driving, this is a satisfying car in most respects, with some imperfections which could so easily be eliminated that their presence is all the more surprising. The style is distinctive, the price buys good quality and the privilege of being different—he who likes this combination will probably become a pleased owner, with the added, comfortable knowledge that this very model is likely to remain in production, unchanged, for many years to come.

Luggage space should be ample; it's provided at both ends.

The interior is roomy enough for two people, but not for four.

Below the rear compartment floor is a 4-cyl, air-cooled engine.

Only heat-resistant luggage should be placed over engine.

ROAD TEST
KARMANN-GHIA VW 1500

SCALE: 10" DIVISIONS

DIMENSIONS

Wheelbase, in	94.5
Tread, f and r	51.6/53.0
Over-all length, in	169
width	63.8
height	52.0
equivalent vol, cu ft	325
Frontal area, sq ft	18.1
Ground clearance, in	5.6
Steering ratio, o/a	n.a.
turns, lock to lock	2.8
turning circle, ft	38
Hip room, front	52.3
Hip room, rear	52.0
Pedal to seat back	39.0
Floor to ground	7.5

CALCULATED DATA

Lb/hp (test wt)	50.5
Cu ft/ton mile	80.1
Mph/1000 rpm (4th)	19.8
Engine revs/mile	3030
Piston travel, ft/mile	1370
Rpm @ 2500 ft/min	5430
equivalent mph	108
R&T wear index	41.5

SPECIFICATIONS

List price	n.a.
Curb weight, lb	1930
Test weight	2270
distribution, %	41/59
Tire size	6.00-15
Brake swept area	128.6
Engine type	4-opp, ohv
Bore & stroke	3.26 x 2.72
Displacement, cc	1493
cu in	91.3
Compression ratio	7.2
Bhp @ rpm	45 @ 3800
equivalent mph	75
Torque, lb-ft	83 @ 2000
equivalent mph	40

GEAR RATIOS

4th (0.89)	3.67
3rd (1.32)	5.45
2nd (2.06)	8.50
1st (3.80)	15.7

SPEEDOMETER ERROR

30 mph	actual, 31.0
60 mph	59.8

PERFORMANCE

Top speed (4th), mph	87
Shifts, rpm-mph	
3rd (5000)	67
2nd (5000)	43
1st (5000)	23

FUEL CONSUMPTION

Normal range, mpg	22/25

ACCELERATION

0-30 mph, sec	5.9
0-40	9.3
0-50	14.6
0-60	21.7
0-70	34.0
0-80	
0-100	
Standing ¼ mile	21.7
speed at end	60

TAPLEY DATA

4th, maximum gradient, %	8.0
3rd	12.1
2nd	17.9
Total drag at 60 mph, lb	100

ENGINE SPEED IN GEARS

4th
3rd
2nd
1st

2000 3000 4000 5000
ENGINE SPEED IN RPM

ACCELERATION & COASTING

90
80
70
60
50
40
30
20
10
MPH

SS¼
4th
3rd
2nd
1st

5 10 15 20 25 30 35 40 45
ELAPSED TIME IN SECONDS

GREEN LIGHT FOR THE
VOLKSWAGEN 1500 K-G

- £1150 tax paid (estimated)
- 84.6 mph. Acceleration: 0-30 6.0 sec, 0-40 8.9 sec, 0-50 14.2 sec, 0-60 20.0 sec, 0-70 29.3 sec
- 27.5 mpg
- 1980 lb
- 169 in long, 64 in wide, 53 in high
- Four-cylinder horizontally-opposed air-cooled ohv engine in rear driving rear wheels; four-speed all-synchro gearbox; 1493 cc developing 53 bhp; drum brakes, 200 sq in lining area; suspension by laminated torsion bars and trailing links (front) and swing axles
- Oil change every 3000 miles; grease at eight points every 3000
- Two/three seat two-door coupe, front and rear boots

GETTING

LADIES AND GENTLEMEN, OUR adversaries:

In the home corner, weighing in at 1918 lb, the German touring defender by Wolfsburg out of Osnabrueck—the VW 1500 Karmann Ghia. Facing it on teutonic soil for this test of power, pep and perseverance, the Dagenham Dandy—at 1968 lb, the Ford Capri 1500. May the best car win.

To labour this prize-ring analogy a moment more, both the Capri and the 1500 KG moved into the litre-and-a-half division from a smaller weight (capacity) class recently and both aim for the same diadem of buyers. The two are strangely similar in a vast number of specifications too—yet different as mild and bitter on the road where it counts. Consider:

Beginning with their common class, two seats apiece and general outlines the cars proceed to be nearly identical in weight though the Ford has a very definite edge in the power/weight race at 33.1 lb/hp against 36.2 for the VW coupe. Though the German car is 50 lb lighter it is 6.5 hp weaker as well. Both engines are fours of course, with Ford championing the in-line water-cooled school and VW sticking to air cooling and a boxer layout. The Capri uses every cranny of its capacity class with 1499 cc and has a good bit higher compression at 8.3 to 1 against 7.2.

Two further engine figures are pertinent to the duo's road behaviour. While the Capri produces its 59.5 hp at 4600 rpm the slower-reving VW gets 53 at 4000, yet outtorques the Ford 83 lb/ft to 81. Incidentally, we are so used to thinking of the new Ford engines as short-stroke units in the extreme it comes as a shock to note that the Ford dimensions are 80.96 bore to

BUSINESSMAN'S EXPRESS
FORD CAPRI

£786 tax paid

86.2 mph. Acceleration: 0-30 5.8 sec, 0-40 9.2 sec, 0-50 13.5 sec, 0-60 19.8 sec, 0-70 28.4 sec

26.5 mpg

2016 lb

Four-cylinder in-line water-cooled ohv engine in front driving rear wheels; five-bearing crankshaft; four-speed all-synchro gearbox; 1498 cc developing 60 bhp; 9½-in disc front brakes, drum rear; suspension by wishbones and telescopic struts (front) and rigid axle with leaf springs

Oil change every 5000 miles, no greasing

Two/three seat two-door coupe, rear boot

PERSONAL

72.7 stroke while the VW runs 83/69.

Finally, before we leave the figures to their separate table, note the identical rear axle ratios. VW achieves further cruising ease by fitting its traditional 'overdrive' top gear, whereas Ford uses a conventional 1 to 1 ratio and accepts the higher revs.

At rest the Volkswagen 1500 KG looks the bigger car, a trick of proportion since it has notably less wheelbase and is slightly shorter overall as well as both lower and narrower —though wider.

To be wholly fair to both cars we'd have to take the 1.5 Ghia to England for a stretch. It grew from a design laid down to compliment the German road—and particularly autobahn—net, and since we tested it almost entirely on its home ground it had every advantage. The Capri still has remnants of British island philosophy and Britain was the one country we didn't take it during comparison testing. Just to put the record straight, though, we do append later UK impressions.

Meanwhile let's examine the two as subjectively as we can. We must mention here that the Volkswagen had all the best of the weather luck while the Capri was subjected to every indignity of snow and ice and very, very sub-freezing temperatures. The fact that it could still rank alongside the VW and score points of its own speaks particularly well for Dagenham. Incidentally, the Capri is the only English Ford handled by the German branch of the common American giant. While they were secretly very unhappy with the older Capri, the Germans perked up considerably with the 1.5. They even preferred it to producing a coupe in their own line.

For two people and travel luggage the pair are very similar.

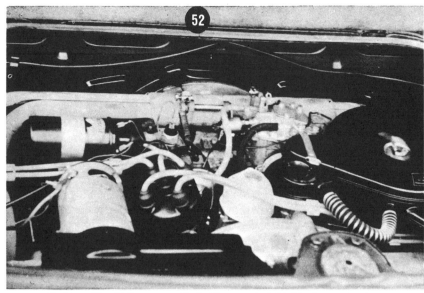

Karmann Ghia's sweeping bonnet has no engine to hide. Instead it reposes, as in the 1500 saloon, under a false floor in the shallow luggage boot at the back

Taking the Volkswagen first, the individual front seats are well-dished with support for both shoulder blades and hips and excellent thigh support. It is no trick to slide right back from the wheel if you care to drive with fairly outstretched arms, and the seat backs adjust—with the qualification that to drive completely in the current GP style you have to recline almost to the point of Lotus. Since you sit well down in the body tub that makes visibility poor.

The Capri has equally fine lateral support but, in true Ford style, you can't really get far enough away from the steering wheel. The other objection is pedal placement. While the passenger enjoys lavish legroom the driver has to sit with his accelerator leg cocked all the time. The pedals are set too far from the toeboard, but they are nicely hung for big feet. Those same size 10s will be in trouble in the VW with its cramped pedal space.

Speaking of seeing out, the Ghia scores on rear vision with a more vertical glass, while the sloping Capri roof renders its rear window at least half unusable. The interior mirror in the Capri is virtually an ornament. To the front and sides the honours are about even, except that the VW can be uncomfortably warm driving into the sun with a windscreen curving into the roof. It is almost too much of a good thing and furthermore there is a gap between the visors and the roof line that seems to catch stray rays.

Side glass in both cars is ample, though the VW loses points for very stiff window cranks, with too many turns to raise or lower the pane. The Capri was also ahead with cranks for the rear side windows as well, against extractor-vent units in the VW. In cold or misty weather the Capri heater could clear its windshield of even ice in a matter of minutes while the VW at full power was much slower.

The German car has just enough heat for all but the coldest days, and the Capri has more than enough even at 20 below zero. The booster fan is noisy in the Ford but that is a small price for such service. Of course the KG sticks with the hundred-turn heater knob between the seats which takes more time than driving if you care to do any fine adjustment. The Capri controls on the dash may be confusing at first but they are quick to use without taking your full attention.

Knob availability is about even in both cars. Those in the VW may all be reached with the seat back and a shoulder belt fastened, while the Capri seat doesn't go back far enough to put them out of range. The same holds true of the shift lever in both cars.

The Volks gearbox is legendary by now and should have stolen that category, but our test car was looser than most (particularly for III to II downshifts) so that the Ford pulled even again. Shorter throws from position to position gave the VW some advantage, but the Ford box has admirable precision. Both cars are up to date in that they boast synchro on bottom gear as well as the other three.

Rear passengers? Not likely in either car, though the VW has a bit more headroom thanks to less top drop-away. The seat back there is very upright and adults will not be comfortable for more than a quarter-hour. On the other hand even 15 minutes with your neck bent to a U in the Ford would be too much. Legroom in both cases is marginal. Call both two-seaters.

The Volkswagen has a rear seat back which folds forward to furnish the chief luggage space. The shelf is neatly padded and fitted with a proper lip to keep small items from slithering out of reach. Like the 1500 VW sedan, the KG has proper luggage boots both front and rear. But they are too shallow to take more than one case each, plus various soft items. Bulky bags go behind the seats. The Capri of course is lavish with its luggage space, to the point where you have to climb in to load up easily. The rear space in the VW does *not* suffer from engine heat.

The matter of exterior styling falls partly to personal taste, but as usual we'll stick our necks out. Our choice is the VW except for the frontal swoops. Both cars nominally carry four lights in front but whereas the Capri has true quad headlights with excellent range, even on dipped beams, the VW really has two headlights and two outsize parking lights. The inside pair are too far from the main lamps to be wired as second headlights, so they call them fog lamps. They are obviously too high to be really useful for that service (so is any legally acceptable fog-lamp, come to that). Really they are victims of the stylist.

On the open road high-speed tracking and steadiness favour the Capri, with its very slight rear weight bias but good general balance. The VW 1500 KG, however, is the best handling VW we have ever driven. It shows far less tail-wag than others in the family, even in strong cross winds. It leans less in the bends too. Steering is about equally precise for both cars.

The VW pulls ahead—literally—in open-road give and take thanks to far better torque. It is a pleasure to use in fast traffic where overtaking times are minimal in fourth, with third gear in reserve for really brisk motoring. The Capri scoots

COMPARING THE SUGAR-DADDY 1500s:

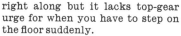

Capri shares many body parts with its sister the Classic, yet as a whole it looks far better balanced. Engine rests between suspension struts amid much space

right along but it lacks top-gear urge for when you have to step on the floor suddenly.

Stopping this momentum (and both cars have roughly equal top speeds in the 80-plus bracket) is a matter of choice. The Capri has discs in front of course and theoretically would be the better car on, say, an Alpine descent. In fact the VW drums proved fully adequate; they didn't fade when a truck with murder in its heart forced a panic stop from 75 mph, which is good enough for us. Using 15-in wheels with correspondingly large brake drums helps here. They are also an improvement in sticky going. The 13-in items on the Ford were hard put to cope with new snow.

The Ford is bouncier on pot-holed roads, though part of its rumble was due to snow treads on the rear wheels. Neither car had excessive body resonance. The VW seemed a touch quieter when driven fast enough to stir up a strong wind.

For cold starts—a Continental measuring stick which will occur to any local buyer after the past winter—the VW has a slight edge but the Capri proved capable of standing overnight at zero and below and starting on the battery every morning but one. That time the thermometer had skidded *up* to −18 after a really bitter night. Starting took a mile and a quarter on a tow rope. The VW was never subjected to that extreme.

The cold undoubtedly had much to do with our exceptionally poor fuel mileage for the Ford. A misfunctioning choke which had to be held in place with one hand for the first 10 minutes every morning also put the consumption up. Still, the VW was definitely more saving of its petrol. Both cars could be fuelled at full flow without blowback but the relatively small Ford tank,

coupled with the high consumption, made the operation necessary far too often for convenience.

Overall, the Ford stood tall to detailed scrutiny but still had the various small bothers of most semi-production or production cars—things like the choke, a restless speedometer needle, an excessively optimistic speedometer and a niggling reverse gear lockout. The Karmann Ghia with about three times the mileage on its clock had virtually no faults. In brief it dis-

played the sort of finish which has made VW famous but which doesn't inevitably appear even on that marque these days.

Which car matches Continental conditions most closely? Despite obvious special efforts by Dagenham, the VW 1500 KG scores for high-speed cruising comfort, trafficability and carefree operation. But the margin is so slim that a bit of applied local knowledge should sell a lot of Fords regardless.

CAPRI COME HOME George Bishop fills in

ON English roads the Capri is rare enough to attract all eyes, yet driving one you spot quite a few brothers about. The driving position is not altogether to our liking; the seat is set so that the driver's right arm is pushed against the door. There seems to be no reason why Ford shouldn't move it in an inch or two.

The steering wheel too is high-set in the Ford manner and has a narrow top rim which tends to make the driver's hands tender on long trips, as on the Cortina.

The driving mirror is almost lethal; it causes a blind spot right in front of the pilot's left eye. The trim round the rear window where the saloon has been cut down to coupe size is shoddy and ill-fitting and not in keeping with the stylish personal-car concept.

But on the road in Britain the Capri is lively enough, unfortunately in more senses than one. There is ample poke and the rear wheels spin readily on loose

stuff, but equally the tail will slide without much provocation.

Whether or not you like its looks comes back to the old tag about beauty being in the eye of the beholder (see pictures). Our car was tomato-red inside and out, not the most restful of colours.

We drove it in some pretty rough weather and it started on the button without a quibble and pulled away well enough from cold. The gearchange was a joy to use and by and large we liked the car apart from the already-listed quibbles about driving position and the skittishness of the tail.

Certainly the Capri is a two-seater—but then that's the idea we suppose, with bags of luggage room for a couple on the move in the behind-the-seats compartment and in the cavernous boot.

Maybe a touring two-seater should be more of a GT model, but Ford has taken care of that by offering just such a beast for those who want it. ●

FORD'S CAPRI AND VW's KARMANN GHIA

For people who can't stand the sight of a Volkswagen.

Some people just can't see a VW.

Even though they admire its attributes, they picture themselves in something fancier.

We sell such a package.

It's called a Karmann Ghia.

The Karmann Ghia is what happened to a Volkswagen when an Italian designer got hold of it.

He didn't design it for mass production, so we wouldn't think of giving it the mass production treatment.

We take time to hand-weld, hand-shape, and hand-smooth the body.

Finally, after 185 men have had a hand in it, the Ghia's body is lowered onto one of those strictly functional chassis.

The kind that comes with VW's big 15-inch wheels, torsion bars, our 4-speed synchro-mesh transmission and that rather famous air-cooled engine.

So that along with its Roman nose and graceful curves, the Ghia has a beauty that is more than skin deep.

USED CARS
ON THE ROAD

No. 241   1962 VOLKSWAGEN KARMANN-GHIA

PRICE: Secondhand £625; New—Basic £871, with tax £1,272

Petrol consumption	30-34 m.p.g.	*Date first registered (in U.K.)*	*13 June 1963*
Oil consumption	negligible	*Mileometer reading*	26,813

VOLKSWAGEN rates of wear and deterioration are still lower than with the majority of comparable cars. Because of this and the fact that the timeless design continues practically unchanged for year after year, secondhand prices remain high. With the Karmann-Ghia Volkswagen, however, demand is not so great, and the used car buyer can have his cake and eat it; he gets the sturdy construction and long life of the VW with a substantial fall in price from first cost as well. With this example, first registered in the U.K. in 1963 (but actually new in Kenya early in 1962), three years have brought the price down to half the original list figure of a new one.

In practically every respect the car is now, if anything, all the better for having covered a "nominal" mileage to loosen and free up all the working parts, and the only marked exceptions to show signs of wear are the brakes. Efficiency is not too good, and roughness when the brakes are used hard at speed suggests that some drum ovality may have developed.

With its well-adjusted automatic choke, the engine is an immediate starter hot or cold, has negligible oil consumption, and also seems appreciably quieter than the average Volkswagen; this may be the result of improved sound insulation in the Karmann-Ghia. The engine does not pink on regular-grade fuel, but does run on badly when very hot. As shown in the table, performance is slow when measured against the watch, with a time of over 50 sec to reach 70 m.p.h. from rest; but the car certainly does not seem sluggish on the road. The somewhat leisurely and "unfussed" sound of the engine encourages one to use to the full what little performance is available, while the quickness of the gear change both saves time and ensures that the gearbox is used freely to keep the revs up. "Instant" synchromesh remains on all four gears, and top gear gives the tireless cruising for which Volkswagens are renowned. Clutch take-up is a little late, suggesting that adjustment is needed.

As well as being quieter than an ordinary Volkswagen saloon, this Karmann-Ghia version also handles and steers better, though there is still the typical strong tendency to wander in cross winds. Oversteer is noticed only in really hard cornering, and is easily controlled. The steering itself is light and has not developed any free play. New Michelin SDS tyres are on the front wheels, and Pirellis (approximately half worn) are at the rear and on the spare.. This mixture seems to work well and not upset the handling.

A curved and much deeper windscreen than the small flat glass screen of standard Volkswagens gives improved visibility, but the driver sits low and sees little of the bonnet and wings. Efficient wipers and the simple but very good pressurized windscreen washer are still working well. The bodywork is in beige below the waistline, with a white roof and surrounds, and only one or two tiny marks and blemishes are present on otherwise excellent paintwork. The chrome, too, has lasted well, except for a few pips of rust on the door handles. There are no rusted edges, and an undersealing compound has looked after the underbody.

Clean Condition

Although the interior finish is not particularly attractive, with its extensive use of plastic, painted metal, rubber floor mats and coarse weave carpet over the backbone, it is certainly durable and conceals signs of use and wear. A few marks are seen here and there, but generally the interior, like the outside, is in very clean condition, even in relation to the comparative youth of a mere three-year-old car. Two small points for attention are that on opening or closing, the engine cover fouls the rear bumper; and the passenger door handle tends to stick.

We set the hands of the clock neatly to 12 after fruitless attempts to get it going; and the only other fault in the equipment is so huge an error in the speedometer that the needle must have slipped. It reads approximately 7 m.p.h. slow throughout the speed range. Four-position adjustment for rake of the front seat backrests is available, and the squab of the rear seat can be dropped flat to serve as a platform for extra luggage.

Lasting impressions of this Volkswagen are of an exceptionally comfortable car, thanks to the excellent ride on its all-independent suspension by transverse torsion bars and trailing arms, and the feeling of complete mechanical reliability. It must certainly have many trouble-free years and miles still ahead of it.

Removal of the radio has left ugly holes in the facia, but the aerial remains with the car. Crazing of the plastic gear lever knob is the only indication of age

PERFORMANCE CHECK

(Figures in brackets are those of the original Road Test, 7 April 1961)

0 to 30 m.p.h.	**6·9** sec (6·6)	Standing quarter-mile	23·8 sec (23·0)
0 to 40 m.p.h.	**11·8** sec (10·9)		
0 to 50 m.p.h.	**19·4** sec (17·1)	20 to 40 m.p.h. (top gear)	15·6 sec (14·4)
0 to 60 m.p.h.	**30·1** sec (26·5)		
0 to 70 m.p.h.	**51·8** sec (—)	30 to 50 m.p.h. (top gear)	16·2 sec (16·0)

Car for sale at: V and F Monaco Motors (Sales) Ltd., 6 Astwood Mews, London .S.W.7. Telephone: FREmantle 4414.

NEXT

As VW sales flounder and sweeping executive changes are made, Bryan Hanrahan tries out the disc-braked 1600 TL Karmann Ghia —and discovers some clues in the desperate search for Clayton's tonic

BEETLE IN DISGUISE?

AUSTRALIA will almost certainly get a version of the "Super Beetle" sometime next year out of the turmoil at Volkswagen's Melbourne headquarters. This will be the old chassis with wider back and front track, modified back suspension, increased glass area, probably the 1584c.c. engine of 65 bhp, and front disc brakes.

The expense of abandoning 1300 and probably 1500c.c. engine versions will be mighty, involving drastic re-organisation of the multi-million dollar expansion plan the company is undertaking to meet the government's 95 percent Australian content requirement.

But the situation is serious enough to demand such measures. In the past 18 months VW's market share has dropped from ten percent to a little over four percent on the latest monthly figures. The smaller BMC cars, the Japanese and the Cortina are the niggers in the Beetle-pile.

Meantime VW are landed with a lot of old stock. Unless the market takes some sort of dramatic upturn there will have to be considerable price cuts to move it.

Just as significant is a recent market preference for the popular medium-size cars. Since late last year the total market for four-cylinder cars has dropped from 34 percent to 25 percent. In the same time the share of the "Big Three" sixes has gone up by eight percent.

What VW needs instanter is a car competitive on the AUSTRALIAN market—not the European, American or South African. The 1300 Beetle is obviously not it.

And the question of a new model, I am told on the best authority, was the reason for the summary takeover of VWA management by German management. The Australians saw the need for a more spectacular car than the 1300: the Germans didn't, at the time.

Now they must face it, and save face at the same time. It shouldn't hurt--financially--too much. They've turned out more millions of Beetles since 1948 than anyone cares to remember, with minimum mods and writing off of dies and machine tools.

Over to Ghia . . .

Having a fair idea of what is coming for us here, this test of the 1600 TL Karmann Ghia with disc brakes is a good pre-sampling.

In the past other model Beetles and Ghias with the same capacity engines have shown fairly similar performances. The Beetle has had the edge on low and medium-speed acceleration because it has been lighter, while the Ghia has rolled out a bit better top speed because of better penetration.

Anyway, the Ghia bodywise is only fractionally different from the last model run, so let's get down to the mechanics and performance.

It is a smooth and satisfying car to drive. But not fast and by the ever-improved standards of today's water-cooled front engines, a bit noisier than it should be.

It is of 1584c.c., four-cylinder, air-cooled, of course, and puts out 65 bhp at 4000 rpm on a compression ratio of 7.7 to 1.

Acceleration from rest to 60 mph is reasonable at 17.2 sec., then after about 70 mph when the traditional VW "overdrive" top gear takes over it quietens down. The figures were: 0-60 mph 17.8 sec., 0-70 27.8. In neutral conditions there is not a lot of urge in the areas where highway overtaking is practised: 40-60 mph in third 12.1 sec., 50-70 in third 12.6 sec., but as with all VWs the slightest help from wind or grade cuts these times considerably.

The car gets all possible help from the four-speed all-synchro gearbox and the magnificent floor change. It is one of the slickest pieces of machinery of its kind yet devised, but whines a bit in the lower ratios.

Fully extended on top speed runs two best times of 94 mph were recorded. The overall average was cut back by a slight head breeze in the other direction to a true 87.2 mph. Once again any little thing that favors the car reflects a lot on performance. Basically the last ten knots or so are a long time coming.

Brakes, handling

The Coronas and bigger-engined Cortinas of today appreciably better the medium and higher speed acceleration figure and run closer to their maximums much quicker.

Whatever I did to the Ghia discs I couldn't make them run ragged. The braking with German Dunlops was as good a disc/drum set-up as I've experienced. Pedal pressures were no higher than with the old drum brakes, in spite of no power assistance.

Re-setting of the back suspension has transformed the handling. The back wheels have slight negative camber in the normal laden attitude. This means less reduction in the track at full downward deflection of the swing axle. The back wheels maintain a wider and more stable platform on the road to resist side forces. Roll effects from wheel tuck-under seem pretty remote even when you're having a real bash.

I really liked the handling. The K-G sat down flat. Steering characteristics are much closer to neutral than before. And, of course, you can't get any steering with a better blend of quickness and accuracy than you do in a VW.

But the back suspension setting in the Ghia is only practical because it is a coupe and is not constantly expected to take weight in the back seats like a Beetle.

Overall fuel figures for 150 miles of testing was 28.8. The tank takes a reasonable 8.8 gallons.

For the rest the Ghia is very nicely put together. The seats have long backs and are comfortable if not sporting: the back seat area is confined and not meant to be permanently inhabited except by the very young — in spirit as well as years.

Instruments are standard VW plus a clock. The only extra equipment (heater and washers are standard) is foglights. The switch is hidden under the dash and they will work only with dipped beam headlight, not on their own. If the foglights are any good they should be better than dipped beams — in which case the dipped beams could often cancel out their effect. A very strange arrangement.

Vision all-round is excellent.

The two doors are very wide. Luggage space back and front is adequate but cramped in shape. No depth.

(Continued on page 63)

SPECIFICATIONS

As for VW 1600 TS fastback except for disc brakes.

PERFORMANCE

CONDITIONS: Fine and warm, two occupants, super fuel.
BEST SPEED: 94 mph.
FLYING ¼-mile average: 87.2 mph.
STANDING ¼-mile average: 21.6 mph.
MAXIMUM in gears: 1st, 29 mph; 2nd, 48; 3rd, 70.
ACCELERATION from rest through gears: 0-30 mph, 4.9s.; 0-40, 8.3; 0-50, 12.1; 0-60, 17.8; 0-70, 27.8; 0-80, 43.1.
ACCELERATION in top (with third, in brackets): 20-40 mph, 6.8s. (5.3); 30-50, 7.2 (5.7); 40-60, 11.9 (9.1); 50-70, 15.0 (12.6); 60-80, 20.2.
BRAKING: 32ft. to stop from 30 mph in neutral; 152ft. to stop from 60 mph in neutral.
FUEL CONSUMPTION: 28.8 mpg over 153 miles, including all tests.
SPEEDOMETER: 2 mph fast at 30; 7 mph fast at 60.

PRICE: $3948 tax paid

Not so much an ultimate in a breed, as an extension of a theme. Meet...

THE KARMANN GHIA 1600 COUPE

ONE of the greatest paradoxes in modern motoring is the NRMA's classification of the VW Karmann Ghia Coupe as a sports car in its current insurance rate-lists. We won't go into how much the NRMA knows about motor vehicles for it has its own reasons for such an unlikely classification, but this does prove one point: that the Karmann Ghia is one of the least understood cars in the world today.

The Karmann Ghia is no sports car — nor does it pretend to be. Actually the car defies evaluation by any existing standards in this country. Just try. Firstly its big asking price (and $3990 is a BIG asking price for this car) puts it into the semi-luxury bracket, but the car's standard presentation and equipment hardly justifies this. Its terminology seems to fit in with a sporting image — but actual performance pulls it right back out of that class. Its seating arrangements make it sound like a small GT — a 2 plus 2 — but no-one considering a Mazda Coupe would have the

same coin to entertain the Karmann Ghia.

No sir. This one is a no-place car that can really only be classed as a highly individual enthusiast's car — which is no classification at all. Its cinderella place in the market is due to certain insurmountable problems.

To start with the Karmann Ghia combined establishments do not allow any old careless hand to assemble their brainchild and this is done under strict supervision in the home country. Secondly they foot VW Germany with a sizeable bill for the privilege of using their coachwork and this doesn't help any when the cars arrive to receive final damnation at the hands of our merciless importation laws. Additionally the cars were originally intended for a selective and small clientele — and never got the benefits of mass production and the corresponding reduction in overall costs, first to factory and then to buyers.

In Australia the Karmann Ghia

is seen so rarely as to be unthought of. But it is the flagship of our local VW establishment and by virtue of its extreme individuality it carries the flag with a certain quiet dignity — it comes as a peace-maker you might say, not a rebel-rouser.

That's the Karmann Ghia in brief, this is how it shapes up in detail—

Take the VW 1600 engine, gearbox, disc brakes and all the other goodies, place a sculptured sporty looking body by Karmann Ghia over it and you have a VW KG. The feature that sets the KG apart from any other 1584 cc Volkswagen (TL Fastback, Twin S sedan or even rally version Beetle) is obviously the body. Now there is a lot to be said for the VW 1300 KG (the old model KG body is still continued using the 1300 cc Beetle engine and is offered in coupe or convertible shape). Its body was clean, unpretentious and with a definite sporting flair. But the Karmann Ghia styling for the 1600 car is a matter of taste. Some womenfolk thought it far more handsome going the other way. We found it looked possible from some angles — overhead three quarters — but just plain ugly from others. But let's not get too personal. The Karmann Ghia attracted plenty of attention wherever we took it.

The first impression the Karmann Ghia leaves is of the customary VW austere quality. The interior reeks of quality control and careful development. It provides full seating for two with an occasional arrangement at the rear preferably for two children. But for two the Karmann Ghia has ample accommodation. The well contoured seats are adjustable fore and aft and for squab position but may be a trifle overfirm. The upholstery of washable vinyl is porous and the perforated centre section breathes well for summer heat. But it is completely

Coming or going? Someone even suggested reversing the design and we can almost see why.

VW KG in the demure surroundings that suits its nature to perfection.

Getting a blurred action shot of the KG is something for it is a docile performer.

unyielding and sometimes uncomfortable.

Cockpit layout is admirably simple with a facia far enough away from the driver's nose not to be like a Volkswagen. Directly in front of the driver and easily seen through the top half of a two spoke wheel is the speedometer with odometer calibrated to whole miles only — but VWs regardless of price have never sported a tenths odometer. Despite calibrations which started at 10 mph and ran to 100 mph! the speedo was pleasingly accurate. Flanking the speedo cowl is a clock (an ideal place for a tacho). On the left another separate cowl houses a half moon fuel gauge and on the lower half warning lights for high beam, fog lamps, oil and ignition. Two switches, lights and wiper/washer on the right and an ill-placed cig-

The VW badge rides high for all in pursuit to see.

arette lighter on the left complete the driver's side. For the passenger there is a facia grab handle, a small unlockable glovebox and an ash tray with flip up ash guard. A VW radio recessed in the centre of the facia produced sound through a tiny two inch speaker which gave a similar effect to the New Vauderville Band. The speaker is cowled similar to the instruments but seems quite out of place.

To accent the sporty flair, the seats are set low which induces the old VW malady of peering over the facia. Certainly the driver has little clue of where the car ends. But you adjust quickly enough.

Driving out into city traffic it was a delight to be using a real gearbox. The KG's gearbox is the best we have experienced in the current range of VWs. Light, fool-proof synchromesh, with a sensibly sized shift knob it is as quick as you like. The clutch unfortunately does not match it. With take up point on the apex of the floor pivoted movement, smooth shifts took concentration. The actual layout is neat with pedals protruding through slots in the flooring. The organ type accelerator was not permanently attached and it was all too easy to catch the accelerator with the shoe welt when transferring from brake to accelerator. This was emphasised by the brake pedal which had an extraordinarily long movement. This pedal arrangement also provided no comfortable resting place for the left foot. Because the seats are low set it is also impossible to bend the left leg and rest it under the right.

Apart from the pedals and over-firm seats the driving position is admirable. The widely adjustable squab and fore aft movement make it easy for any size driver to find the right relationship to the wheel and gearshift. Certainly the fashionable stretched arm position is possible unlike some real sporty cars.

Performance never has been a VW strongpoint. Despite initial comments from two staff drivers who thought the KG was quite nippy, the acceleration figures are not brilliant.

The 1584 cc engine is the same twin carburetted version used in the 1600 TL or the Australian TS Fastback. It follows the VW tradition of being understressed and long wearing.

(Continued on page 62)

Engine fits tidily under well-insulated floorpan in rear compartment. It doesn't transmit noise to the interior.

Combining front and rear compartments, and occasional room from folded-down back seat the Ghia has very acceptable luggage space.

Ghia's styling could best be termed distinctive, or unusual, for few found it really attractive.

Interior comfort is excellent — as you might expect for the price. Equipment is fairly comprehensive.

TECHNICAL DETAILS

MAKE .... Volkswagen
BODY TYPE ...... coupe
OPTIONS .... nil
MODEL ...... Karmann
Ghia 1600

PRICE $3990
COLOR ...... two tone,
black, white
WEIGHT ...... 17.8 cwt

MILEAGE:
Start 1063   Finish 1459

FUEL CONSUMPTION:
Overall .... 28.2 mpg
Cruising .... 29-34 mpg

TEST CONDITIONS:
Weather .... fine
Surface: hot mix bitumen.   Load: two persons
Fuel .... premium grade

PERFORMANCE

Piston speed at max bhp .... 2120 ft/min
Top gear mph per 1000 rpm .... 19.8 mph
Engine rpm at max speed .... 4300 rpm
Engine rpm at cruising speed .... 3600 rpm
Lbs (laden) per gross bhp (power to weight) 35

MAXIMUM SPEEDS:
Fastest run .... 95 mph
Average of all runs .... 86.9 mph
Speedometer indication fastest run .... 97 mph
In gears:
1st 28 mph;   2nd 50 mph;   3rd 75 mph;   4th 86 mph

ACCELERATION:
0-30 mph .... 5.8 secs
0-40 mph .... 9.6 secs
0-50 mph .... 14.4 secs
0-60 mph .... 22.1 secs
0-70 mph .... 30.1 secs

	3rd gear	4th gear
20-40 mph	11.3 secs	NA
30-50 mph	9.2 secs	14.4 secs
40-60 mph	10.2 secs	13.5 secs
50-70 mph	15.6 secs	13.8 secs

SPEEDOMETER ERROR:

Indicated mph	30	40	50	60	70
Actual	26.7	38.2	51.2	60.5	69.5

STANDING QUARTER MILE:
Fastest run .... 20.5 secs
Average of all runs .... 20.7 secs

ENGINE:
Cylinders .... four, horizontally opposed
Bore and stroke .... 85.5 mm by 69 mm
Cubic capacity .... 1584 cc
Compression ratio .... 7.5 to 1
Valves .... overhead, pushrod
Carburettors .... twin downdraught
Power at rpm .... 65 bhp at 4600 rpm
Torque at rpm .... 87 lb/ft at 2800 rpm

SPECIFICATIONS

TRANSMISSION:
Type .... four speed manual
Clutch .... single dry plate
Gear lever location .... central floor
Ratio: 1st 3.08; 2nd 2.06; 3rd 1.32; 4th 0.89. Final
drive 4.125 to 1

CHASSIS and RUNNING GEAR:
Construction .... central tube platform chassis

SUSPENSION:
Front .... trailing arms, torsion bars, a/roll bar
Rear .... trailing arms, torsion bars, swing axles
Shock absorbers .... telescopic

STEERING:
Type .... worm and roller
Turns 1 to 1 .... 3.5
Turning circle .... 36 ft

BRAKES:
Type .... disc front, drum rear
Friction area .... 82.2 sq in.

DIMENSIONS:
Wheelbase .... 7 ft 10.5 in.
Track, front .... 51.6 in.
Track, rear .... 53 in.
Length .... 14 ft 0.5 in.
Height .... 4 ft 4.6 in.
Width .... 5 ft 3.8 in.
Fuel tank capacity .... 8.8 gals

TYRES:
Size .... 6.00 by 15
Make on test car .... Dunlop B7+

GROUND CLEARANCE:
Registered .... 5.4 in.

Graph: ACCELERATION THROUGH GEARS WITH CHANGE POINTS

3RD 75 MPH
STANDING ¼ MILE 20·7
2ND 50 MPH
1ST 28 MPH
TOP SPEED 86·9 MPH

MPH / ELAPSED TIME IN SECONDS

KARMANN GHIA

(Continued from page 60)

But a little more power would be more in order especially when its 1500/1600 cc contemporaries produce from 75 to 85 bhp in standard single carburettor form. Because it is air cooled it suffers the disadvantage of greater engine noise than a water jacketed engine but the effect together with exhaust roar sounds very Porsche-ish which appeals according to age. Over 400 miles of testing and hard driving the KG returned 28 mpg, a satisfactory figure for a 1600 cc, 18 cwt car but light cruising could produce up to 40 mpg.

HANDLING — A FULL TIME JOB

Types of driving suit types of handling. The enthusiastic VW driver either adores or abhors the VWs oversteer. Initial city and inter-urban hopping made us think the KG was to be the least oversteering VW

to date. This was mostly due to initial understeer at low to medium speeds. But driven hard at high speeds the understeer is only momentary and pronounced oversteer takes control. Even one staffman who is used to VWs and rear engined cars found the KG a particular handful both on dry bitumen and gravel at speed. In the wet caution was automatic. Despite the cars quick steering it was hard to make use of the oversteer for quick motoring over indifferent surfaces, a characteristic that the experienced VW driver finds virtuous. This more than usual tail happy handling is due to the low down rearward weight bias slung between swinging rear axles. The natural movement of the wheels when braking or cornering is to swing down and cause the car to run on the side of the tyre area available, giving far less traction.

A comment on one staffman's report suggested the steering wheel should be smaller in diameter with reduced gearing for less effort. As is, the steering is light at speed but heavy at town speeds and for parking. VW has an excellent steering set up but over recent years the factory reached an unfortunate compromise in trying to reduce road reaction. Increasing the damper effect has made the steering heavier than on VWs of the late 50s but it has not in turn eliminated all reaction. A number of times on dirt when recovering from slides, harsh thumps were felt through the steering. All up we feel the steering made lighter would improve the car.

But a very bright feature is the braking. Employing

the disc front, drum rear arrangement the stopping power is superb with an excellent handbrake thrown in. It is an ill-advised move to use heavy brake applications in cornering, but in one emergency situation we found the car totally stable. For normal driving they are faultless and refused to fade on test.

Although the actual action of the brakes was fine we found the pedal movement increased greatly during the test. Considering the discs are self-adjusting and there was no fluid loss we were unable to explain this. Nor was VWA or any of the dealers. With the greater travel the problem of hooking the right foot under the accelerator pedal when transferring from brake to accelerator became most annoying.

Another point that seemed out of character was the ride on anything but very smooth surfaces. Again the conventional VW set up of torsion bars front and rear plus anti-roll bar at the front gave the typical VW harshness. But you can't win them all. The torsion bar suspension is rugged and takes our beaten tracks without murmuring, and after all the car is not intended as a bush buggy.

THE GRUMP DEPT

We had some minor grouches with the car. Our particular car was white with a black top — a little unpractical for an Australian summer. All white would have been much better. Along the same lines there is still no floor or facia fresh air ventilation as with all Volkswagens. The heater has controls on the tunnel near the handbrake and these are simple, easy to use and produce geysers of hot air from floor vents which can be individually controlled. Demisting is controlled by two levers, one for each side, which direct fresh air onto the screen. We can only assume these work as 80 degrees and 20 percent humidity are not conducive to misting. Although the controls are few they all had their idiosyncrasies. We searched exhaustively for a switch to control the fog lamps which are set in the sculptured flairs at the front. We never did find the switch and the handbook didn't help any but it appears that a certain sequence of turning on the main lights and flicking the dipper switch will bring them on. The dipper switch is a key bar incorporated in the inside of the turn indicator arm, similar to the German 1600 TL Fastback. When the lights are off, the same key operates as headlight flasher for day use. If the key is held in the lights automatically flash in sequence with the turn

"Who needs overdrive?"

indicators. With all lights and ignition off the turn indicator controls side parking lights separately on each side. For parking in one way streets at night this admirable idea should be more widely used. Warning lights are provided for both high beam (blue) and the fog lamps (green). They are of the fanned shutter type used in the 1500/1600 series VW and ideal as warning lights.

The standard seat belts are of the two point sash type. We felt that in a really quick stop they would have strangled the occupants very satisfactorily. There was a great tendency when braking hard for the body to slide forward on the shiny seats underneath the straps.

Luggage space is adequate. There is a compartment in the nose and one under the rear engine cover (a lined floor sits over the engine and is removed with twist rings). Careful thought has put an extension dip stick protruding through the trailing edge of the "boot" so the floor does not have to be removed to check the oil. In the nose there is enough room for a large suitcase and a couple of soft overnight bags. When two are travelling in the car the rear seat can be laid flat and the whole rear compartment loaded. Using all compartments there is adequate room for holiday luggage but getting it in and out proves a bugbear that every VW owner has to put up with.

At night the headlights are not adequate for the ouring speed (75-80 mph) of the Karmann Ghia. On ʌow beam a very short chopped off beam spreads light immediately in front of the car and this is helped in intensity if one is lucky enough to fluke switching the foglamps on. On high beam the typically low powered VW lights are lost all too soon. When do we see a VW with a 12 volt ignition?

The ignition cum steering lock mounted on the steering column had a sensible but annoying arrangement: after a mis-start it is impossible to recrank the engine without switching off the ignition and starting afresh. The Volkswagen 1600 Karmann Ghia falls into a category apart from all other cars — similar to the Austin Vanden Plas 1100. Taken for what they provide in motoring, cars like this are quite acceptable. Set to an Australian price tag they become market misfits, with a restricted, selective clientele. The VW has quality — the carpets fit exactly, the doors close with a quality thunk and there are no left over bits of gum on the trim and so on. It also has a narrow field of application in our motoring scene. As far as its owners will be concerned this is probably a good thing, after all, it was not designed for plebeian tastes.   #

BEETLE IN DISGUISE

(Continued from page 57)

One thing the Ghia has that the Beetle hasn't—and in Australia is so important — is fresh-air ventilation through the scuttle. It does not admit enough air really, but some is far better than none as anyone who has been confined in a Beetle on a hot day well knows.

The other thing VW should look at in their re-organisation is prices of parts and service. It's all very well to have fixed prices, provided they are fixed at the right level. Some major VW replacement assemblies like engines and transmissions cost more than twice Holden replacements, as a glance at any lists will show. And what position is the factory in to do anything about this now that it has closed so many dealerships? Certainly the problem is going to be more difficult.

But at least if the super Beetle comes there could be one basic engine for all cars and commercials, which is the sort of rationalisation that any motor car maker goes for and could help the customer's pocket.

There's also in Europe a long-rumored 1.8-litre car and the Audi subsidiary which makes a front-wheel drive car.

My information is that neither of these will be the schwerpunkt (spearhead, to you) of VW's new effort in Australia.   ●

An Electric Volkswagen Karmann Ghia

Built as a commuting car using parts available now

AN electrically powered commuter passenger car can be built today with available hardware. It will meet the limited performance characteristics necessary and perform comparably with an internal combustion engine-powered vehicle for the same application.

This is the contention of two engineers from the Allis-Chalmers Corporation's Advanced Technology Centre in West Allis, Wisconsin, USA, after long-term experience with an electrically powered Volkswagen Karmann Ghia. They are Tom M. Thiele, electronic and control technology director, and David L. Moore, unit manager, electronics and control technology.

The vehicle, operated by Mr Thiele, carried him between his home in Oconomowoc, Wisconsin, and Allis-Chalmers, a total of 27 miles each way. In a 10-month period ending at the time of the report, the vehicle travelled almost 10,000 miles, most of it on Interstate Highway 94 and the business streets of West Allis.

The car can operate for one hour between charges while travelling at 60 mph. However, for city driving at lower speeds, considerably greater range is possible.

"With the 27 h.p. direct current, shunt-wound type motor used, total alternating current power consumption from a standard 220V circuit averaged one-half kilowatt hour per mile travelled," the two engineers report.

"The car demonstrated a number of important things. Reliable transportation was provided throughout the year, including days of −20 deg. F. with no vehicle breakdowns. Although a maximum distance range of 65 miles at 60 mph was demonstrated, variations in headwind, temperature and travel requirements made it advantageous to have recharge facilities at both ends of the route." These facilities were in Mr Thiele's garage and at an Allis-Chalmers laboratory.

The report also said that the car logged about 200 miles per gallon of battery fill, which is distilled water. Wisconsin's cold winters require a gas-type car heater, although waste engine heat defrosts the windscreen.

For purposes of cost identification and comparison, the researchers assembled actual cost data on two similar Karmann Ghias with conventional engines. One was purchased new and driven 124,500 miles in six and a half years, the other purchased used and driven 48,300 miles in five years. Actual per mile costs were 5.06 cents (5p) for the new car and 6.77 (about 7p) for the used car, reflecting such costs as licence fees, initial cost less selling price, maintenance, and repair, gasoline and insurance. Adjusted for today's prices, the costs became 6.16 cents (6p) and 7.52 cents (7½p) respectively.

Using cost figures developed with the electric powered version and including a projected energy tax of 1.3 cents/kwh, the researchers were able to project an overall cost comparison for the two engine-powered autos if they had been orignally electrically powered. Thus, per mile costs for the new car would be 7.48 cents (7½p), a difference of plus 21.4 per cent, and for the used car 9.13 cents (9p), up 23.8 per cent over the vehicles' costs as engine-powered machines.

A number of factors would make electric vehicle costs competitive with engine-powered designs, the researchers said. As background, they noted that utilization of vehicles with modest performance characteristics already is high in the US; in 1969, of all cars sold in the US, 12.9 per cent were imports many of which have modest characteristics. Many families can use modest performance cars as their second and third cars; 27 per cent of families owning cars in 1969 had two or more.

The compact arrangement of motor and control in the Allis-Chalmers electric passenger car (top) is made possible by the field control on the shunt-wound traction motor used in conjunction with a standard clutch and transmission. The front compartment (bottom) is not so battery-loaded that the spare tyre is crowded

By George E. Toles

Factors that would make electrics competitive include: Streamlined design for bodies made of non-deteriorating material, such as tough plastics. This would reduce depreciation costs and so reduce total costs, and also lessen aerodynamic losses and so increase available power; special power rates for electric car owners to encourage off-peak hour battery recharging as a means to creating profitable power load for generating stations; additional research in battery construction and composition to help to reduce weight and space requirements.

In terms of vehicle design, the factors to be considered, are: Vehicle configuration must provide sturdy construction to support a maximum battery weight/total vehicle weight ratio while maintaining lowest possible aerodynamic and rolling friction losses; The battery system must provide maximum energy density at high discharge rates with the lowest total cost per mile; The power trains and controls must be designed for maximum energy conversion efficiency. In addition, a simple design will enhance reliability and improve market acceptance.

Specifications are given for an electric commuter car based on today's available hardware, Mr Thiele and Mr Moore said. Such hardware has been used in the test vehicle, has been tested for potential use in it, or has been studied while in use in comparable situations.

The Karmann Ghia used for experimental work is practical for only two passengers, although a larger version could be built to carry four, according to Mr Thiele. In the test Karmann Ghia, the rear seat area was filled with batteries. The balance of a total of 30 12V batteries weighing 1530 pounds was placed under the front bonnet.

Low losses permitted by an enclosed underside and streamlined contour give an aerodynamic drag coefficient of 0.3. This reduction, plus the effect of low-loss radial-ply tyres, produces a requirement of only 13.3 h.p. at 60 mph. Some further reduction in losses could be obtained from low-loss compound tyres, which are not available in this vehicle's size.

In order to prepare the VW Karmann Ghia to carry the battery weight, the rear torsion bars were replaced with others capable of supporting 3600 pounds including a 175 pound driver. Heavy duty batteries were used.

Special motive power 12V lead-acid batteries were selected as the best compromise between low cost and energy density available. The batteries were connected in series-parallel to supply 120V.

The control system involves an efficient direct current drive configuration in which a dc shunt wound motor with field control is coupled through a standard transmission and clutch to the differential and driving axle. Speed is controlled by a low power dc "chopper" (this is an electronic power input controlling device) which manages the field current in response to the accelerator. The arrangement eliminates need for a baulky, expensive chopper in the armature □

37 ▶

weather, but believe it should be quite snug—at least at cruising speeds—and the rear window defroster is a valuable addition.

DON'T DO-IT-YOURSELF

Engine accessibility is good; you open the lid and there it is. But the VW engine is rather like an iceberg—the bulk of it is submerged beneath sheetmetal. This built-in deterrent to do-it-yourselfing helps account for the booming after-sales service dollars earned by the dealers. You can easily fiddle with the carburetor, the points and the oil filter—and you can change your own plugs with some difficulty (Hazet, a German firm, makes a very good spark plug wrench just for VW and Porsches). Aside from that, routine maintenance only consists of changing the oil and adjusting the valves. Neither of these is particularly difficult, it's just that it's so much easier to drive in and have it done. The battery is in plain view on the right side of the engine so topping up is easy. The brake master cylinder in the trunk is a translucent plastic so the level can be checked every time you fill the gas tank.

Once you find everything, servicing a VW is not difficult. The factory supplies a very detailed owner's manual (but just try and get a shop manual) which helps quite a bit, but it soundly discourages home service whenever avoidable.

The trunk lid opens high and the opening itself is very wide. It will carry enough baggage for two for an extended trip and its space can be supplemented by the behind-the-seats platform. The rear seat back folds down to provide a capacious space for luggage. It continues quite deep under the rear deck all the way to the firewall. The spare is located in the trunk to help offset the engine weight and to place it where it can be checked regularly. The high lip of the trunk opening, however, may discourage some women from attempting a tire change since the tire must, of course, be lifted clear of this. On this one feature at least the Dauphine seems to offer a better solution. The very forward end of the trunk, in addition to housing the spare and windshield washer tank, also holds the jack and handle neatly clipped in place, and the tool bag and spare fan belt all dry and clean.

We drove the Ghia well over 1000 miles, achieving an overall gas mileage of 33 mpg. As we said, no oil was used and we had no trouble with the car other than the low brake pedal. Three of us took a weekend trip to Watkins Glen for the SCCA races, a total distance of over 500 miles and found that this number of adults could easily be accommodated since the one located on the "emergency seat" could sprawl comfortably. Two children would fit nicely back there and for carrying infants, the fold-down seat back creates a good-sized area.

SERVICE IS THE BEST POLICY

The importance of good dealerships and readily available, uniform service is as big a part of VW's story as the car. No other imported car has made such strides in this area. Heinz Nordhoff, President of Volkswagen, said, "The availability of first-class service everywhere has played a key role in Volkswagen's success. This view is no mere partisanship—anything but." He continued, ". . . service has become a science of its own. We have such a lead in this field that none of our challengers can catch us if we continue on our present path." Mr. Haan, commenting on VW dealers' successes, said "successful

dealers are nothing more than alert, intelligent and fair-minded businessmen."

In order to keep the old customers happy and the new ones arriving in increasing numbers every year, Volkswagen of America has a dynamic, many-faceted policy. One aspect of this is its Model Dealership Planning Guide, containing plans and specifications for eight dealership layouts ranging from 3864 to 27,000 square feet for aiding new dealerships in getting set up. A typical dealership would be built on a 2.3 acre tract and contain a six-car showroom area, a fully-enclosed reception area, a 3400 square foot parts department, and an 8600 square foot service department. It could have a separate body and painting department. Its outward appearance would be similar to one in almost any major population area you could name. It would be modern and distinctive—immediately recognizable as a VW agency. No other cars would be sold there except VWs and it would probably be the only one in the area.

As is apparent from the specifications, parts and service are not neglected, in fact these areas account for most of the space in a dealership. Modern communications methods assure dealers of having an adequate supply of parts. Orders from dealers and distributors are sent to Volkswagen of America and are filed on punch cards. These are sorted and transcribed to tape. The tapes are sent to Wolfsburg where they are "played" and the orders are filled.

With its eyes wide open, Volkswagen is launching an assault on the fiercest competition in the U.S. sales battle. There are 19 imported and 36 U.S. models in its price range, these figures including most of the compacts and smaller sedans. However, the future looks bright; as things are now, VOA claims the Karmann-Ghia is "the second best selling two-seater in the United States, second only to one domestic car" (Corvette).

Volkswagen has never made any claims that the Ghia is a sports car—and neither would we. It is sporty, however, and should appeal to a great number of people seeking pleasurable driving in an attractive package. It's more than just transportation, but offers all the owner benefits that the standard VW does. The standard equipment, for example, includes the heater and defrosters, directional signals, wheel trim rings, an electric clock and the padded dash. For 1962 the cars also have seat belt brackets built-in for the front seats. The quality control exercised in producing the cars is extraordinary and production details are thorough. These include mounting all the fuses under the dashboard in a clear plastic box where they can be easily reached. The finish of the parts you seldom see seems equal to the exterior; the underdash area is evenly painted. Carpeting and rubber mats complete the trim. Two door pockets supplement the glove compartment. An assist grip is built-in for the passenger. The passenger's toeboard is angled for more comfort and a "courtesy light" is standard. The seats are well-made, wider and more dished than the standard car's.

The Ghia is Volkswagen's concept of Grand Touring for two. It's made for people possessing a sophisticated automotive design sense (both esthetic and technical), people who want to go places quickly and efficiently in an atmosphere of sporty good taste at a realistic fee.
—C/D

'

You have to admit
the Speedster and
the Ghia kind of look
alike, and they *feel* alike,
and they perform
the same. But more
importantly, they even
handle the same—
and handling was what
the Speedster was
really all about
when it came down to
the final analysis.

'

The Last Speedster Ever Made Is... Ya Gotta be Kiddin'... A Karmann Ghia'?'

BY BOB BROWN

No car ever inspired such fierce loyalty as the Porsche Speedster, and no Speedster owner ever looked on a Karmann Ghia but with utter disdain. Well, look again, Marque Chauvinist Pigs, a Ghia just went by you.

• No one who ever owned a Porsche Speedster escaped coming down with a dose of Marque Chauvinism. To the Speedster freak, there simply was no other car . . . a Healey was little more than a rude assemblage of engineering principles better left to agricultural applications . . . a Triumph was the same, but more emphatically so . . . an Alfa came closer, but all its sophistication had been lavished on the engine—and it was overly temperamental for that . . . a Corvette was patently a case of Detroit-inspired cubic inch overkill. Plus they all shared one Great Drawback—the engine was stuck up over the steering wheels, a long driveshaft away from the driving wheels. How rudimentary. How uninspired. How . . . quaint.

For *every* other car made, the Speedsterphile had only varying degrees of disdain. *Every* car . . . even the Porsche 356 Coupe from which the Speedster had sprung was not exempt—it was a "nice" car, but with its watertight fast-back roof and heavily padded seats, it was relegated to the faint-praise category. It was condescendingly written off as "more of a *touring* car than a true sports car." A little too . . . uh . . . all-purpose, if you *know* what I mean.

But while the Speedster owner merely disdained all other cars, there was *one* in particular that was held in blackest contempt. Not just because of what it was—which was a standard VW Beetle trying to pass itself off as a sports car—but because of what too many people thought it was—which was a Speedster being sold by VW. The *Karmann Ghia!*

Foul poser. Pretender! Defiler!

It was very much an emotional hatred, because, in general, the Speedster had an uncomfortable similarity to the VW.

After all, it was rear-engined, it had that small light, quick steering feel. It had even been designed by Porsche himself. It certainly was no "Porsche," but it was undeniably closer than anything else on the road. And even if they didn't actually like the car, collectively the Speedster clan knew more about it than any comparable group of VW owners. They had to because of a fact that was kept well suppressed to all but insiders . . . there were a hell of a lot of VW parts that could be used as low-cost replacements for Speedster parts. Almost the day the first Speedster arrived, an underground telegraph sprang up to spread the word that, say, a VW transporter clutch would bolt right in at a third of the cost of a replacement clutch bearing the "Porsche" stamp. Or in a pinch, a $17.00 VW steering box could be used in place of the $80.00 real thing. Maybe that's why all Porsche owners of the time got slightly shifty-eyed when someone would innocently ask, "A Porsche? That's the Volkswagen sports car, isn't it?"

"No! It isn't! The Karmann Ghia is a VW!" would be the foam-flecked denial—even if the denier had just spent an afternoon hanging over a VW service desk to read the parts manual upside down.

The worst part about this strain of Marque Chauvinism is that it is chronic. As with malaria, time and medication can lessen the severity of the attacks, but no one ever completely shakes it. (See Steve Smith's accompanying autobiography if you doubt this fact.) Even now, 13 years after the last Speedsters rolled off the line, former owners recall their Speedsters as the most enjoyable car they have ever owned . . . and still

detest the Karmann Ghia.

All right, all you guys, we'll get it over with right away . . . the 1972 Volkswagen Karmann Ghia is the Speedster re-incarnate. It does nearly everything the original did just as well—and many things it does better.

And don't think that we came by that decision easily . . . several members of this staff are former Speedster junkies who have had to face up to the fact too. And the fact is: All of a sudden, after all those years, the goddamn Karmann Ghia doesn't seem so bad.

Look at it! After 17 years in production, the Ghia has become such a familiar sight that hardly anybody ever really *looks* at it anymore. But when you do, you see that its once galling similarity of silhouette is as close as any 1972 car dares come to the simple sheetmetal drape that Porsche laid over its 356 chassis to create the discount-price Speedster (sold in 1956 for $2995). Even with the addition of Department of Transportation-mandated chunky taillights and sidemarkers, the Ghia, like the Speedster, gives the impression that it is wearing a size-too-small body . . . that the sheetmetal had to be stretched to cover the skeleton underneath. The result is an appreciated air of bulbous functionality rather than the extremes of neo-Thirties automotive classicism or ersatz Group 7 sculpturing that predominates elsewhere today.

And better than just being visually reminiscent of the Speedster, the Ghia feels similar to sit in. And you do sit *in* it, which, as any Speedster owner will admit, was one of that car's greatest attractions. When you got into a Speedster you sank damn near out of sight. The car was low enough to begin with (most magazines of the time reported its overall height as 51 inches . . . actual height was three inches less), but it felt even lower because of the very high door sills that meant you stepped over and *down* into the car. And once the door shut behind you, you found yourself eye level with a seven-inch high windshield and your shoulder just about level with the scantily cushioned top of the door. There was never any doubt that you were in the car—it was almost impossible to avoid the feeling that you were in command of a very mobile pillbox from which you could view the world perfectly over the twin bulges of the front fenders, while still remaining almost invisible yourself. (The feeling of isolation naturally became even more pronounced with the top in place, and may have more than a little to do with the Speedster legend . . . there are all kinds of stories of what went on inside top-up Speedsters parked on busy streets that have absolutely nothing to do with any of their intrinsic qualities as cars.)

The Ghia creates very much the same environment, even though it comes equipped with a full-size windshield. You twist sideways through the narrow door openings and drop into the seat, and suddenly it's 1956 all over again. There's that same giant steering wheel an arm's length away and a vista of sheetmetal rolling away from you in all directions. Even the wand-like shift lever is there with its long third-gear throw maybe three-quarters of an inch too far way. So too are the tightly grouped foot controls that allow you to accomplish the heel-and-toe maneuvers required of all *sports car* drivers. And from behind you comes the only slightly subdued sound of a small air-cooled motor, thrashing and whirring—occasionally throwing in a syncopated clacking. It's all so familiar . . . but something just isn't right. *It's the seat.* It's one of those VW straight-back chairs, mounted on hijacker seat rails that's throwing everything off. But even without going to new seats, it's possible to drop the OEM seats almost an inch and give the seat back some additional rake with a drill, hacksaw and some wooden blocks. *Then* try to say it's different.

About all you can point to is the dashboard, and there we'll have to agree. VW doesn't even supply a speedometer marked off with shif' points anymore, not to mention a tach. How the hell could any Speedster owner drive a car so remiss in supplying basic information? What happens when the clutch cable breaks (which it did about every 850

miles on every Speedster ever built)? With a tach all that would mean is the inconvenience of a push start until you got around to threading a new cable . . . but to be asked to risk a gearbox solely on the sensitivity of your ear is too much even for a Speedster driver. On the real thing the only gauge you could see without shifting hands was the tach—and for that you barely had to shift your eyes. Searching through the local Pep Boys for a tach is going to be required, if a Ghia is to be seriously considered as an ersatz speedster, only if the Ghia does more than *feel* like a Speedster. It has to do more than that—a lot more. It has to perform the same, certainly, but most importantly, it has to handle—*that* was what a Speedster really was all about.

When it came time for the head-to-head confrontation at New York National Speedway, the staff of *Car and Driver* had laid enough personal bets to have been able to open one of New York's Off Track Betting offices right on the sixth floor of One Park Avenue . . . and almost every one of them on a purely emotional basis as to what every bettor *wanted* to happen. We've tested VWs; we've tested, driven and owned Porsches, but we must have contracted a collective case of Speedster Marque Chauvinism because we've *never* tested a Karmann Ghia. (In fact, it was hard to find a staff member who would admit to having driven one in the past 18 months.) And the outrage of the premise we were testing out was seemingly verified just in acquiring proper test cars. We had expected trouble in finding a 1600 Normal Speedster that would be representative of the over-the-counter version sold in 1956-58, but we finally located Don Yuhas, about 22 miles from our office. His car, aside from a rollbar and an Abarth exhaust (both of which he felt necessary for his occasional participation in autocrosses) was strictly as sold in 1956, for $2995. For its rival, we selected a Karmann Ghia Convertible (what else?) with no options, which currently has a suggested list price of $3099—certainly a reasonable cost-of-living increase over more than a decade, should it match up.

It does. Despite its rated 10 horsepower deficit to the Speedster, smog gear and 130 extra pounds, the '72 Ghia didn't let down its backers by much in the acceleration tests (where the Speedster group was expecting a TKO). It proved to be just over a second and 4 mph slower than the Speedster at the end of the quarter-mile. (To prove that we weren't using a Speedster with a wet

fuse, Yuhas' car ran 18.6 sec. which compares more than favorably with contemporary road tests of the car in 1956 which claimed 18.8 for a 1600N and 19.2 for one equipped with a 1500cc engine.) We would be less than candid however, if we didn't state that the Porsche certainly felt—and sounded—better. Part of the difference undoubtedly came from the VW's puny one-barrel carburetor and exhaust system that requires spent gases to travel through a Victorian maze before exiting the car through two chromed pipe stems, but at least as much of the lack of performance feel is the fault of the transmission. The Porsche smoothly climbs its way through its four gear ranges with no drastic drop offs. In undeniable contrast, the Ghia rears up on the line (there is virtually no anti-squat in the suspension) and sails up to top rpm in first and second nearly as quickly as you can shift. But there is a tremendous gap between second and third (see data panel) that ruins it all. It's like you've driven into an invisible wall of elastic Saran-Wrap and the Ghia has to struggle forward. Unfortunately, it never bursts through that boundary, as fourth—as in the Speedster an overdrive range—is more of the same. The answer to this problem is getting better over-3000 rpm performance from the VW engine—not just for extra performance but to increase drivability. The solution isn't free, but it's readily available from one of the multitude of manufacturers specializing in VW speed equipment—like carburetor/manifold kits and extractor exhaust systems and—Speedster fans will particularly love this—different transmission ratios.

In the all-important handling test the Ghia devastated the Speedster. It was able, even on its standard-issue tires, to duplicate the Speedster exactly. VW's double jointed rear axles and semi-trailing arm suspension (adopted in 1968) are what make the difference. It works *better* than the Speedster's rear suspension. On the skidpad both cars generated identical 0.73 G lateral acceleration—with the Ghia exhibiting *less* understeer than the Speedster below that maximum, and neither prone to making the transition to oversteer without giving forewarning. (Tire technology has certainly improved the Speedster in this aspect.) But the VW uses a slower steering ratio along with more conservative spring rates and, consequently, has an unwanted softer feel with more body roll than the Porsche. Where this was particularly noticeable was on the handling loop rather than the skidpad. Here the

1972 Karmann Ghia	**1956 Porsche Speedster**
PRICE Base Karmann Ghia convertible $3099	$2995

ENGINE

1972 Karmann Ghia	1956 Porsche Speedster
Type: Flat Four, air cooled, magnesium crankcases, aluminum heads	Flat Four, air cooled, magnesium crankcases, aluminum heads
Displacement.........................96.7 cu in, 1584 cc	96.5 cu in, 1582 cc
Bore x stroke.............3.36 x 2.72 in, 85.4 x 69.1 mm	3.25 x 2.91 in, 82.5 x 74.0 mm
Compression ratio...........................7.3 to one	7.5 to one
Carburetion..............................1 x 1-bbl Solex	2 x 2-bbl Solex
Valve gear..............Pushrod operated overhead valves	Pushrod operated overhead valves
Power...................60 bhp @ 4400 rpm (SAE gross)	70 bhp @ 4500 rpm (SAE gross)
Torque...............81.7 lb-ft @ 3000 rpm (SAE gross)	82 lb-ft @ 2700 rpm (SAE gross)
Maximum recommended engine speed.........4400 rpm	4500 rpm

DIMENSIONS

1972 Karmann Ghia	1956 Porsche Speedster
Wheelbase.....................................94.5 in	82.5 in
Length..164.0 in	155.2 in
Width..64.3 in	65.5 in
Height...52.0 in	47.9 in
Track.....................................51.6/53.3 in	51.3/49.9 in
Curb weight..................................1980 lbs	1850 lbs
Weight distribution F/R.....................41.7/58.3%	43.2/56.8%

DRIVE TRAIN

	1972 Karmann Ghia		1956 Porsche Speedster	
Transmission	...............4-speed manual, all synchro		4-speed manual, all synchro	
Final drive ratio	.......4.13 to one		4.28 to one	
Gear	Ratio	Mph/1000 rpm	Ratio	Mph/1000 rpm
I	3.80	4.6	3.09	5.5
II	2.06	8.5	1.76	9.7
III	1.26	13.8	1.23	13.8
IV	0.89	19.6	0.89	19.2

SUSPENSION

1972 Karmann Ghia	1956 Porsche Speedster
Front: Ind., trailing arms, transverse torsion bars, anti-sway bar	Ind., trailing arms, transverse torsion bars, anti-sway bar
Rear: Ind., semi-trailing arms, transverse torsion bars	Ind., swing axle, trailing arms, transverse torsion bars

STEERING

1972 Karmann Ghia	1956 Porsche Speedster
Type.................................Worm and roller	Worm and roller
Turns lock-to-lock...................................2.9	2.0

BRAKES

1972 Karmann Ghia	1956 Porsche Speedster
Front:..10.9-in disc	1.57 x 11.0-in drum
Rear:................................1.57 x 9.0-in drum	1.57 x 11.0-in drum

TIRES AND WHEELS

1972 Karmann Ghia	1956 Porsche Speedster
Wheels.................4.5 x 15-in, stamped steel, 4-bolt	4.5 x 15-in, stamped steel, 5-bolt
Tires.....................Continental 5.60-15, bias ply	Continental 165HR-15, radial ply

PERFORMANCE

1972 Karmann Ghia	1956 Porsche Speedster
Standing ¼-mile...................19.7 sec @ 67.4 mph	18.6 sec @ 71.8 mph
70-0 mph braking.......................190 ft (0.86 G)	220 ft (0.73 G)
Lateral acceleration (200 ft skidpad)..............0.73 G	0.73 G

Ghia's softness would allow the inside power-transmitting wheel to go light on acute turns and its skinny tires were simply not enough to permit power to be fed back in smoothly. That was probably the essence of the difference—the degree of smoothness with which the cars could be driven hard. In terms of absolute handling performance, the cars were very nearly equal—even to the extent of sounding alike as both produce the same reassuring whonks and thumps as the suspension works to keep the wheels on the pavement—but the Speedster does it without straining while the VW seems a bit unfamiliar to its role as a sports car.

However, under braking, just the opposite is true. It is the Ghia which stops as if it were second nature, while the Porsche is uncertain. And in this case the Ghia not only feels better, the test results show that it was significantly more capable. The VW would consistently generate perfectly controllable 0.86 G stops from 70 mph while the best the Speedster could do was 0.73 G. Moreover, as the Porsche's all-drum brakes heated up on repeated stops, it became increasingly difficult to prevent lock-up and subsequent surrender of directional stability.

The damn Karmann Ghia. Once again it's a thorn under the saddle of every Speedster fan. And this time 'round it may even be worse. Right now the Ghia sells for just over $100 more than a new Speedster did. But there *aren't* any more new Speedsters. The Karmann Ghia is the closest thing to the genuine article that you can find—only if you lust after what the Speedster represented, rather than what it did, will you be uneasy. Not only that, but there are enough aftermarket parts available (including quick-steering kits and stiffer shocks) to make the Ghia's performance *and feel* identical with a good Speedster's—and they're cheap enough to do it for well under the price of restoring a decade old car that has probably been raced (and certainly has rusted out around the fender liners and bellypan). Even if the Karmann Ghia isn't sold in silver gray, or ivory white or that instant-oxidizing blue or red that the Speedsters wore for identification, and even if it means you'll have to live with roll-up windows and a watertight padded top that doesn't drum on your skull when it's in place, it isn't that much of a sacrifice. You just have to accept the fact that the last Speedster built is a Karmann Ghia. ●

KARMANN GHIA

Volkswagen inside, Italian styling on the outside.

In the ever ensuing argument of what is a sports car, the VW based Karmann Ghia has its advocates. While we have come to think of sports cars as high powered vehicles, we often overlook the fact that the old classic TC's were not exactly stormers as to horsepower. The basic VW engine and platform chassis, along with the simple and reliable suspension system has been used as a starting point for everything from dune buggies to sleek fiberglass bodied GT coupes. In the case of the Karmann Ghia, the factory did the same thing. They had the Italian stylist Ghia, design the body which is built by Karmann, the German custom body builder who does bodies for Porsche. Two models are available, a hardtop and a convertible with a well padded and practically flapless top that looks almost like a solid one when it is up.

Previous models of this semi-custom VW had a kind of rear seat back that folded to a flat luggage carrier, but this has now been eliminated in favor of a fully upholstered cargo area. The front bucket type seats have fully adjustable backs and there is plenty of leg and head room for the 2-passenger capacity. Like the Beetle, air conditioning and a choice of dress-up options are optionally available for this little sportster which has the same power plant as the Type 1's.

Because of the availability of high performance VW engine goodies, the Karmann Ghia is a favorite for the hop-up fans, with bigger barrels, hot cams, dual carburetion, and extractor exhaust systems giving them the go-power to stay with cars of higher power. Because of the low 7.5:1 compression ratio, these little engines respond well to su-

percharging by either the engine driven or turbo exhaust types.

The flat four, a horizontally opposed and air cooled engine, has a displacement of 96.66 cubic inches and is rated at 46 SAE net horsepower at 4000 rpm, not wheel spinning power but enough to allow comfortable all-day cruising at 60 to 75 miles per hour, with a possible top of about 90 mph. Because of the better aerodynamic shape and about 50 or so pounds less weight than the Super Beetle, slightly better performance can be expected unless one is hauling a pair of 250 pound passengers. Included in VW's computer analysis program is the Ghia, which is also wired with the plug-in connection for the service checks now being installed in major Volkswagen dealerships.

New for this little 2-seater for 1973 are an improved air intake heater for cold starts and stronger bumpers as required under the latest regulations. Dual exhausts, beefed up transmission synchros and a diaphragm clutch for the 4-speed manual transmission are all part of the continuing hidden changes for the Ghia. In addition, caliper disc front brakes combined with drum type in the rear are standard and for those who like to shift but detest the clutch action required, the optional automatic stick shift is available. For '73, this option now has a park position on the selector that allows stopping while in gear without stalling the engine.

As mentioned earlier, this is not a muscle car, but the fully independent rear suspension that several years ago replaced the old swing axle system, plus the reliable double trailing arm front end, gives the Ghia a lot of fun driving stability. Unitized construction, improved engine mounts, and added insulation, provide a quiet ride with minimum transfer of road and engine noise

to the passenger compartment. The Ghia has the same 24 month, 24,000 mile warranty as the other VW machines, with free computer diagnosis during the warranty period.

With the growing swing toward only hardtop configurations, the convertible model may be among the few such soft tops available for very much longer. Domestic makers have just about phased out their "ragtops" but a few importers are still supplying some for those who like to feel the outside while riding inside.

KARMANN GHIA
Data in Brief

DIMENSIONS

Wheelbase	94.5 in.
Overall length	165 in.
Height	52 in.
Width	64.3 in.
Tread — Front/rear	51.3/52.7 in.
Steering type & ratio	Worm & roller
Fuel capacity	10.6 gal
Luggage capacity	3.7 cu ft front
	14 cu ft rear
Design passenger load	2 passenger
Turning diameter	36.9 ft
Curb weight	1918 lbs

ENGINE — Standard

Type	OHV 4-cylinder, opposed air cooled
Displacement	96.66 cu in.
Horsepower	46 at 4000 rpm
Torque	72 lbs/ft at 2800 rpm

DRIVELINE

Transmission	4-speed all synchro manual (auto stick-shift opt.)
Drive axle ratio	3.875:1
	4.125:1 (convert)

BRAKES

Front	Caliper disc
Rear	drum

SUSPENSION

Front	Independent, double trailing arms, torsion bars, tubular shocks
Rear	Independent, single trailing arms, torsion bars, tubular shocks

WHEELS & TIRES

Wheels, type & size	Steel, 4½ J x 15
Tires, type & size	6.00 x 15

NA — Data not available
DNA — Data not applicable

FOR RICHER VOLKS
A NEW WAGEN

A low silhouette (above) and sharply creased nose (left) emphasize the pert good looks of the Ghia-Karmann V.W.

One of the hottest current news items for those interested in the People's Car is that Volkswagen has a glamour model! But before anyone protests that they don't want a change, let it be emphasized that the car pictured on this page is an *addition* to the line, and the old familiar "beetles" will continue to purr from Wolfsburg, beeping at each other on the highways of the world (albeit with raised tail lights and built-in blinkers).

Ghia has designed the new body, and it is being built by Karmann of Germany. Although a stock engine and standard VW chassis components are used, the smooth contours of the new body will permit a top speed about 10 mph greater than the sedan. Designed as a two-seater coupe, the car has an overall length of 163 in., a width of 64.2 in., and is 52.2 in. high. Ground clearance is 6.8 in., and unladen weight is 1782 lbs. There is 6.4 cu ft. of space behind the seats and 2.8 cu. ft. in the front compartment.

The Ghia-Karmann VW will make its U.S. appearance in late September and the western selling price will be about $2450. ●

This two-seater coupe body offers 6½ cu. ft. of space behind seats. Except for narrow grille, there is no evidence of engine location.

True beauty has been achieved on the short wheelbase by Ghia of Italy, through skilful designing. Seating has been lowered in order to lower centre of gravity.

How the new Ghia-Karmann coupe compares.

Ghia-Karmann versus Volkswagen

Ghia looks better, handles better, outshines the

Volkswagen on the road.

MOST VW owners, to put it mildly, are car-proud folk, and will Go On for hours on end about their beetle-like cars. Their flat four aircooled engines, it would seem, are virtually unbreakable. The independent torsion bar ride gives the VW a leech-like roadability. The solid German engineering is a miracle of precision. The little beast will scale hillsides like a kangaroo, tuck itself into unparkable corners, run on the open road for miles and miles as fleetly as a scared rabbit, and over all this, it will turn in a m.p.g. figure that is gratifying, to say the least.

So they say . . .

The same owners, however, when asked a straightforward question like "How do you like the body styling?" are apt to either burble platitudinous words like "functional", and "utilitarian" — or else greet you with a heap of enthusiastic silence.

Especially the latter, should the owner be a woman.

That's where the Ghia-Karmann coupe comes in — or would, were it possible to import it into Australia.

Volkswagen themselves long ago realised the photographic shortcomings of their little car, and fairly recently decided to do something about it. (See "Wheels", Feb. '56). With the help of Carozzeria Ghia in Turin, Italy, VW's Wilhelm Karmann was turned loose with a free hand to beautify the beast — and did so.

Huge Hearted Willingness

Result is a good-looking, aerodynamic steel and glass coupe which has beneath its sleek exterior all the huge hearted willingness of the traditional Volks. Minor chassis changes were made, such as the slight modifications to the steering gear and front suspension, but although a slight difference to carburetion has been made, mechanically both cars are identical.

The question was, how would they compare on the road; a question which "Wheels", with the aid of overseas correspondents, will now try to answer:

Improved Handling and Roadability

The Ghia coupe will out-corner the Volkswagen any day of the week, and produce less side sway in so doing. Obviously this is due to Karmann's lowering of the car's centre of gravity by the simple expedient of dropping the seats to a lower position within the frame. The steering column angle has been lowered accordingly, and a stabiliser added to the front suspension. Roll angle, as a point of interest, is 2 degrees for the Ghia as compared with 3 degrees in the case of the VW.

Fuel Consumption Tests

TABLE A—GASOLINE MILEAGE FROM PERFORMANCE TESTS

M.P.H.	Ghia	Volkswagen
20	52.2	49.5
30	50.2	46.5
40	45.0	42.0
50	42.2	36.3
60	35.0	30.0
City-Traffic	26.8	26.9

table:

ACCELERATION—GHIA vs. VOLKSWAGEN

M.P.H.	Ghia	Volkswagen
0-20	4.08	4.20
0-30	7.68	8.15
0-40	13.00	13.7
0-50	20.1	22.5
0-60	34.2	45.0
20-40	10.8	10.1
20-60	32.0	41.4

Best top speeds obtained from both cars, incidentally, were Volks, 67 m.p.h.; Ghia-Karmann coupe, 71 m.p.h.

Braking Tests

As regards braking ability, the Ghia stays right alongside its stablemate and goes to the head of the class. In a dozen consecutive stops on both cars, brake fade was not apparent on either, but pedal pressures were considerably less on the Ghia, probably due to softer linings combined with the car's lighter weight.

The Ghia-Karmann offers cold-climate purchasers a new type of heater as an optional extra.

Small Back Seat

Some things on the Ghia, however, have not changed. The four cylinder aircooled engine, which, incidentally, turns out 36 b.h.p. as does its counterpart, still howls like a banshee during acceleration. The low speed roughness is still there, and headroom in the back seat is even worse than in that offered by the Volkswagen. Still, one must grant certain concessions to the stylist, and the wonder is that the Italian designers have managed to fit a back seat in at all.

Both interior and exterior finish in the Ghia-Karmann is of excellent quality. Chrome work has been kept to a strict minimum, but hardly appears to be necessary, such is the beauty of the new coupe's line.

At present, import restrictions prevent the shipment of these attractive little cars to Australia, however, should it be possible to bring them in at a later date they can be certain of meeting a substantial public demand, particularly since, as a VW product, they are certain to receive the same enlightened and comprehensive brand of world-wide service which has already become a Volkswagen byword.

At time of writing, one Ghia Karmann coupe is in Australia. It is the personal conveyance of Dr. Blank, a German Embassy diplomat stationed in Canberra. Certainly it is an attractive little vehicle, which attracts the attention of a crowd wherever it goes.

"Wheels" has been given to understand that these cute little tricksters are not being imported here because the conditions appertaining to VW's import quota are based on all vehicles being brought in c.k.d. (completely knocked down).

Assembly of the Volkswagen is therefore done entirely in this country.

The Ghia, on the other hand, having a coachbuilt body, would need to be imported fully assembled, and this would necessitate another special import license. ●

Cockpit of the Ghia is roomy, well laid out without being ostentatious. Steering column has greater rake than on VW.

New Ghia-Karmann coupe has trim lines, larger luggage space.

Country	38.1	35.7
Overall	28.1	23.8

The table above tells a pretty graphic story, and shows that under all conditions excepting those of city traffic the Ghia uses less fuel than the Volkswagen. Taken in comparison with the acceleration tests it is likely that the Ghia's relatively poor economy in traffic (though comparable with the VW) is due to a somewhat richer acceleration mixture coming through the modified carby. City conditions, with constant use of the accelerator being required, would accentuate this. However, all other types of driving showed that the Ghia-Karmann coupe preferred to take its nourishment in sips, rather than the medium-sized gulps preferred by the Volkswagen.

Aerodynamic

How then, has Karmann whipped better gas mileage *and* acceleration, together with a higher top speed, out of a machine which is fundamentally identical to its beetle-browed counterpart. The answer may be found in the aero-dynamics chamber, where it has long been known that good streamlining and lessening of a car's frontal area gives rise to less horsepower being needed in order to push the car along. The Ghia proves this point to everybody's satisfaction. Compare the fuel economy figures given, with the following acceleration

New Beutler-Volkswagen Coupé offers style and luxury

BEAUTY in automobiles must be paid for. Individuality—the glow you get when your car is unique—also costs money. When the two appear in one car like the Beutler Special Coupé the price is high—around Sfr. 17,000 (£1,390 approx.) ex Thun, Switzerland, for the "standard" model.

Yet the best recommendation for the style, workmanship and finish of this Volkswagen variation is the fact that people will still reach for their wallets after hearing the total. This VW Special, produced by the Beutler Brothers near Bern, will be the sire of a "private series", trimmed in each case to suit the customer. They'll even alter wing or roof lines slightly if you wish.

I referred to the "standard" model because it is hard to say what the basic price includes. Luxury touches like a "light" horn, two-way mirror, cigar lighter and washer, padded visors with a vanity mirror on the right, Ghia speedometer, clock and oil and fuel dials, and a grab handle for the passenger are just some of the "standard" fittings.

Extra cost options on the test car—which Ernst Beutler uses for private transport—include twin fog lights, Porsche

Nose view of the Beutler Special Coupé, to use the full name. The fog lights are not standard

It looks just as fine from the tail—an unusually well-balanced car

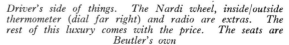

Driver's side of things. The Nardi wheel, inside/outside thermometer (dial far right) and radio are extras. The rest of this luxury comes with the price. The seats are Beutler's own

wheels and brake drums, inside/outside thermometer, sliding roof, radio and Nardi steering wheel, as well as the twin carburetter Beutler manifold.

The Beutler body is mounted on an absolutely stock VW chassis which explains part of the cost. They have to buy finished cars and scrap the beetle bodies. The test machine was on the latest chassis with roll bar in the front and it handled very much like a normal Volkswagen—only more so.

The extra 110 lb. coupled with a lower centre of gravity and slightly better weight distribution, gave the car a steadier feel in the fast curves but it cries for more push. The Special was fitted with the two Solex 32 PBIC carburetters I mentioned but no other tuning. This option

The rear seat folds down to reveal a cubby for small items or to make more luggage space. As a seat, it falls into the "occasional" category. Detail work is evident here too

(running some Sfr. 800) should just about balance out the added weight on acceleration but the sleeker form would support Beutler claims of an 80-85 m.p.h. top speed. The car was too new to test fully.

The driving position contributes a great deal to the sporting feel of this charmer. The seats are a Beutler design and they give thigh and back support where it should be. The cloth panels in the middle help the semi-bucket design in holding you in place. They slide well back for straight-arm driving.

The shortened gear lever, right where your hand falls, operates the standard VW gearbox which is fine for the car. If anything, the shorter lever makes changes even crisper than normal.

Visibility from a Beutler is excellent. Both front wings are up where an average driver can see them without stretching and the wide rear window—with a low engine deck—gives an abundance of rear view.

I don't suppose any car is perfect, but the negative marks I must give the Beutler are all items that could be easily altered with the special-order nature of the car. The worst was a finish of shiny black enamel for the top of the dash. It picks up reflections from overhead and can be very distracting.

Of lesser importance was the lack of identification on switches and the fact that the ignition key got in the way of the turn indicator hand. Also, just to be fussy, I would prefer a tachometer to the clock. Lastly, the wipers on the left-hand-drive model cleared plenty of glass for the driver but left a blind spot in front of the passenger.

In the rear the Beutlers have padded and upholstered the occasional bench seat fully, but it is still too shallow for adults going any distance. With the front seats well back it is extremely shy of leg room, too. More to the point, the back folds forward to make a wide, handy luggage area.

Incidentally, the nose compartment itself is one of the biggest surprises of the Beutler. By VW standards it is vast and, more important, of a very usable shape. The fuel tank and spare tyre lay flat under the boot mat, leaving an oblong compartment that will hold a great deal.

At the rear the engine is housed in a roomy compartment with a lid that you control from inside the car. On the negative side it is placed somewhat deeper and further forward than on a standard car, to the slight detriment of service.

Finish on the Beutler Special was absolutely beyond reproach. Not only did all seams and joints look beautiful, but the doors and lids all closed with that light pressure and pleasant snick that indicates quality fitting. Every panel on the car is hand-fitted, which helps to explain

the 12 weeks the two Beutlers and their crew need to produce one.

This is admittedly no car for the economy minded but the price is fully justified by the quality. And the design is one of those rare birds that looks just right. Simply expressed: it's a very, very hard car to walk away from.

Ernst Beutler, one of the brothers in the firm's title, holds back the boot mat to reveal the flat spare tyre. The filler cap is just visible ahead of the tyre and the tools are in a box behind it. The space is vast for a VW and very useable in shape

Business end with the engine somewhat deep for easy access. The optional twin carburetter unit from Beutler was fitted to this car

modern MOTOR ROAD REPORT

RICH FOLKS' WAGON

Natty Karmann-Ghia coupe clothes standard Volkswagen components in new style and luxury — and now it can be bought right here, reports David McKay

UNTIL recently there were probably less than half a dozen Karmann-Ghias in Australia—Sydney had three, Canberra one, and that was the lot, as far as I know.

Wherever these shapely cars went they attracted attention and gave rise to mild speculation on a new VW. The bold VW hubcaps were easily recognisable in passing, and the KG was soon labelled "the new VW." But, as time went by and nothing radically new came from Wolfsburg, the rumors began to fade.

Then suddenly, in November, Lanock Motors (the N.S.W. VW agents) shook the local VW world: on their showroom floor appeared a shiny new Karmann-Ghia convertible.

That first evening William Street was full of VW's braking to a halt, doors flying open and VW addicts rushing to peer through the plate-glass at what was surely a new model at last!

But next day inquirers learnt the truth—yes, the KG was for sale (at £1850, including radio, screen-washers and seat covers)—and no, it was NOT a new VW model.

Progressive N.S.W. manager Doug Donaldson had ordered a couple of KG's despite VW Australia's doubts. The rush of inquiries has stimulated business and justified Donaldson's action. VW orders needed this shot in the arm, it seems, for even the addicts were starting to yearn for some changes to their beloved

"beetle"—if only to show that they owned the latest model!

Wolfsburg has answered their prayers to a certain extent—as you'll learn from Bryan Hanrahan's road test, elsewhere in this issue, the latest VW has received many worthwhile modifications. The engine is quieter, the rear suspension far less frisky; there's a new dished steering wheel, self-cancelling trafficators, and a stabiliser bar on the front end.

Now all these new features of the standard VW are also found in the KG, which in itself has been much improved since the first model was released at the 1955 Frankfurt Show.

Being among the hundreds who visited Lanock's showroom to inspect and rhapsodise over the KG con-

Behind the two seats is an occasional seat for children, the back of which folds down to disclose a very useful luggage compartment. This occasional seat can be folded flat, so that a really generous amount of luggage can be carried when travelling two-up. Then, of course, there's a bit more space in the front "boot" alongside the spare wheel and fuel tank.

At the rear of the car is found the faithful air-cooled VW engine— standard in every respect and developing the same 36 b.h.p. at 3700 r.p.m. Alongside the engine is mounted the 6-volt battery, as in the "commercial" VW's. It leads a rather warm life in here and needs frequent topping-up.

Chassis and suspension are straight-out VW, but the lovely KG body is decidedly heavier than the VW shell, and weight distribution differs from that of the standard car. The KG scales another 1½ cwt. — and although the inevitable penalty is paid in acceleration, the KG wins hands-down on roadability.

CONTROLS and instruments are straight-out VW, including the new dished wheel; but cockpit layout is classier, pillars slimmer.

vertible, I was able to wheedle Donaldson's own KG away from him for a road test.

He warned me to expect shouted queries from passing traffic, particularly in the Harbor Bridge queues. He also said the price for the hard-top test car was £1598—and that he had an embarrassingly long waiting list for the possible extra dozen or so that might come in during the year, all subject to the approval of the Wolfsburg factory and VW Australia, of course.

The test car was a lovely sight in its orange and off-white color scheme; fully imported, it had a Porsche-like quality about the finish, both inside and out. All that seemed to be missing, apart from the Porsche's more energetic engine, were the fully reclining seats.

The KG has superb seating, but the seat-backs have only the normal VW three positions. Yet, behind the wheel, you get that Porsche feeling again—that the car is built around **you** and for you **personally.**

Everything is at hand—gear lever and handbrake perfectly placed, the foot controls somewhat more offset than in the VW, but perfectly spaced for the easiest of all "heeling-and-toeing" movements.

ENGINE is standard VW, developing 36 b.h.p. at 3700 r.p.m. It has to push 1½ cwt. more, but better roadholding offsets loss in acceleration.

Instrumentation is basic, in traditional VW fashion. There is the normal VDO speedo, plus warning lights, and a matching clock in place of the Porsche's rev-counter. Between these two large dials nestles a fuel gauge. Windscreen-washers are fitted, and the test car had an AWA radio of good tone.

Gone are the restful sun-visors that VW owners have appreciated for years; instead, in keeping with modern safety policy, you get two padded visors—effective but not as restful.

On The Road

First thing I noticed about the KG was its silence—gone is the usual VW whine, which has worried back-seat passengers in the past. Next I found the car anything but sluggish —it seemed to revel in traffic-light G.P.'s, and I had to watch those red lines on the speedo or the engine would spin silently and quickly well past the indicated change-points.

The ride over some of our worst city roads was very Porsche-like; there was little trace of the rather short, choppy movement of the VW—altogether far smoother, yet without any sloppiness or "float."

But the real revelation was to come in the handling department. It was now all Porsche and no VW. Certainly, there was a certain degree of oversteer — but it was very controllable and never unpredictable.

Out on the road, I must admit to being gratified by the admiring glances of young and old alike. I can't recall ever driving a car that attracted so much attention.

In the KG the VW owner gets the lot. He has been able to praise his "beetle" to all and sundry, but he's never been told his car was "lovely." The KG is just that—and it has the wonderful advantage of combining good looks with reliability.

Perhaps the only thing lacking is a little more performance. Here, I think, the Swiss M.A.G. blower is the answer. Any home mechanic can fit the unit in a couple of hours, and the improvement is astonishing.

Compared with a standard 1958 VW, the KG averaged 47 m.p.h. over

MAIN SPECIFICATIONS

ENGINE: 4-cylinder, air-cooled, horizontally-opposed, o.h.v. bore 77mm., stroke 64mm., capacity 1192 c.c.; compression ratio 6.6 to 1; maximum h.p. (S.A.E.) 36 at 3700 r.p.m.; Solex downdraught carburettor, mechanical fuel pump, 6v. ignition.

TRANSMISSION: Single dry-plate clutch; 4-speed gearbox synchromeshed on top three; overall ratios, 1st 3.60, 2nd 1.88, 3rd 1.23, top 0.82 to 1; reverse 4.63 to 1; spiral bevel final drive, 4.4 to 1 ratio.

SUSPENSION: Front independent, by trailing links, laminated torsion bars and stabiliser bar; divided axle and torsion bars at rear; telescopic shock-absorbers all round.

STEERING: Worm-gear type; 2¾ turns lock-to-lock, 36ft. turning circle.

WHEELS: Pressed-steel discs, with 5.60 by 15in. tyres.

BRAKES: Hydraulic, 96.1 sq. in. lining area.

DIMENSIONS: Wheelbase, 7ft. 10½in.; track, front 4ft. 3½in., rear 4ft. 1¼in.; length 13ft. 7in., width 5ft. 4½in., height 4ft. 4¼in.; ground clearance, 6½in.

KERB WEIGHT: 16cwt.

FUEL TANK: 8.8 gallons.

PERFORMANCE ON TEST

CONDITIONS: Fine, warm, no wind; smooth bitumen, two occupants; premium fuel.

BEST SPEED: 80 m.p.h.

STANDING quarter-mile: 23s.

FLYING quarter-mile: 75 m.p.h.

MAXIMUM in indirect gears: 1st, 20 m.p.h.; 2nd, 45; 3rd, 70.

ACCELERATION from rest through gears: 0-30, 6s.; 0-40, 10s.; 0-50, 15s.; 0-60, 22s.

ACCELERATION in third: 20-40, 8.6s.; 30-50, 10.0s.

ACCELERATION in second: 10-30, 5.0s.; 20-40, 6.6s.

BRAKING: 30.5ft. to stop from 30 m.p.h.; 34.3ft. after fade test.

FUEL CONSUMPTION: 30 m.p.g. overall for hard-driven test.

PRICE: £1598 including tax

FRONT "boot" contains spare wheel and fuel tank, leaving only modest space for occupants' luggage . . . *BUT there's a deal more room behind the rear seat—and with the backrest down ample luggage can be carried inside.*

my regular mountain circuit and climbed the test hill in 2 min. 53 sec., while the VW recorded 46 m.p.h. and 2 min. 58 sec. Yet the VW was 0.5 sec. faster from 0 to 30 m.p.h. and a whole 2 sec. faster to 50!

The KG was better at holding its maximum speed — here weight and shape helped—and its better roadability offset the VW's nippiness.

The test KG was a far nicer car than the one I drove in Germany in 1955. It seems safe to say that KG policy will be like VW policy—continually improving a winning design.

VW
Karmann-Ghia

1200 Karmann-Ghia Convertible

ENGINE CAPACITY 72.90 cu in, 1,192 cu cm
FUEL CONSUMPTION 37.6 m/imp gal, 31.4 m/US gal, 7.5 l x 100 km
SEATS 2 + 2 **MAX SPEED** 74.5 mph, 120 km/h
PRICE list £ 945, total £ 1,142

ENGINE rear, 4 stroke; cylinders: 4, horizontally opposed; bore and stroke: 3.03 × 2.52 in, 77 × 64 mm; engine capacity: 72.90 cu in, 1,192 cu cm; compression ratio: 7; max power (SAE): 41.5 hp at 3,900 rpm; max torque (SAE): 65 lb/ft, 9 kg/m at 2,400 rpm; max number of engine rpm: 4,500; specific power: 34.8 hp/l; cylinder block: cast iron; cylinder head: light alloy; crankshaft bearings: 3; valves: 2 per cylinder, overhead, push-rods and rockers; camshaft: 1, central, lower; lubrication: gear pump, filter in sump, oil cooler; lubricating system capacity: 4.40

imp pt, 5.28 US pt, 2.5 l; carburation: 1 Solex 28 PICT downdraught twin barrel carburettor; fuel feed: mechanical pump; cooling system: air-cooled.

TRANSMISSION driving wheels: rear; clutch: single dry plate; gearbox: mechanical; gears: 4 + reverse; synchromesh gears: I, II, III, IV; gearbox ratios: I 3.80, II 2.06, III 1.32, IV 0.89, rev 3.88; gear lever: central; final drive: spiral bevel; axle ratio: 4.375.

CHASSIS backbone platform; front suspension: independent, twin swinging longitudinal trailing arms, transverse laminated torsion bars, anti-roll bar, telescopic dampers; rear suspension: independent, swinging semi-axles, swinging longitudinal trailing arms, transverse torsion bars, telescopic dampers.

STEERING worm and roller, telescopic damper; turns of steering wheel lock to lock: 2.40.

BRAKES drum; braking surface: total 96.12 sq in, 620 sq cm.

ELECTRICAL EQUIPMENT voltage: 6 V; battery: 66 Ah; dynamo: 180 W; ignition distributor: Bosch; headlights: 2.

DIMENSIONS AND WEIGHT wheel base: 94.49 in, 2,400 mm; front track: 51.38 in, 1,305 mm; rear track: 50.71 in, 1,288 mm; overall length: 162.99 in, 4,140 mm; overall width: 64.33 in, 1,634 mm; overall height: 52.36 in, 1,330 mm; ground clearance: 5.98 in, 152 mm; dry weight: 1,808 lb, 820 kg; distribution of weight: 43% front axle, 57% rear axle; turning circle (between walls): 37.7 ft, 11.5 m; width of rims: 4''; tyres: 5.60 × 15; fuel tank capacity: 8.8 imp gal, 10.6 US gal, 40 l.

BODY convertible; doors: 2; seats: 2 + 2; front seats: separate, adjustable backrests.

PERFORMANCE max speeds: 24.2 mph, 39 km/h in 1st gear; 46 mph, 74 km/h in 2nd gear; 63.4 mph, 102 km/h in 3rd gear; 74.5 mph, 120 km/h in 4th gear; power-weight ratio: 43.7 lb/hp, 19.8 kg/hp; carrying capacity: 794 lb, 360 kg; max gradient in 1st gear: 41%; acceleration: standing ¼ mile 22.9 sec, 0 — 50 mph (0 — 80 km/h) 17.1 sec; speed in top at 1,000 rpm: 18.3 mph, 29.5 km/h.

PRACTICAL INSTRUCTIONS fuel: 85 oct petrol; engine sump oil: 4.40 imp pt, 5.28 US pt, 2.5 l, SAE 10W-20 (winter) 20W-30 (summer), change every 3,100 miles, 5,000 km; gearbox and final drive oil: 5.28 imp pt, 6.34 US pt, 3 l, SAE 90, change every 15,500 miles, 25,000 km; steering box oil: 0.18 imp pt, 0.21 US pt, 0.1 l, SAE 90; greasing: every 1,600 miles, 2,500 km, 12 points; sparking plug type: 175°; tappet clearances: inlet 0.008 in, 0.20 mm, exhaust 0.012 in, 0.30 mm; valve timing: inlet opens 6° before tdc and closes 35° 5' after bdc, exhaust opens 42° 5' before bdc and closes 3° after tdc; tyre pressure (medium load): front 16 psi, 1.1 atm, rear 20 psi, 1.4 atm.

VARIATIONS AND OPTIONAL ACCESSORIES Saxomat automatic clutch.

VW with yankee punch

Bud Hopkins takes his sneaky Volkswagen Karmann-Ghia out for a stroll in the afternoon. Beware all you big bore drag racers, here is a little German import that might thrash your big iron. You see, Bud has a new Corvair engine in the rear.

Putting ein grosse Corvair engine mit der kleine Volkswagen is making mach schnell rennenwagen

Text and photos by John O'Donnell

If you were to take a poll of the world's motoring public, you'd probably find that the most popular of the really small automobiles is the Volkswagen. But if looks alone were considered, then the outcome would undoubtedly be different. Of course, a significant deviation of this point is the Karmann-Ghia. The Ghia has clean functional lines that have been a source of admiration for several years. But then the problem of snappy performance crops up. Compared to the average American car, the aforementioned little jobs just plain won't go the way we'd like them to.

Bud Hopkins of Albany, California has done something to remedy this sit-

uation. Along with Claude Warren and the Lukes & Shorman Machine Works, he has transformed his VW into a top rate screamer. By installing a stock 80 horsepower Corvair engine he has more than doubled the original output of his 1958 Karmann-Ghia. With a weight increase of only 45 pounds, the handling characteristics have not been adversely changed. Hopkins feels that it has improved due to the increased weight giving the rear wheels a little more negative camber.

The actual installation of a Corvair engine into the Volkswagen Karmann-Ghia is not difficult. It does, however, require the facilities of a machine shop.

In the standard Volkswagen sedan the rear body lines would have to be altered to clear the engine. In the Karmann-Ghia and the VW Commercial the body lines are not affected.

The first step in the conversion begins with the Corvair engine. Hopkins and Warren dismantled the engine and removed the crankshaft. Using a 1961 VW Commercial flywheel as a pattern, they machined the end of the Corvair crankshaft so as to fit into the VW flywheel end boss. Again using the VW flywheel as a guide, they drilled the ends of the Corvair crankshaft so that the VW locating pins are used to position the flywheel to the Corvair crank.

The flywheel end of a Corvair crankshaft is recessed, so this recess must be tapped to take the VW pilot bearing bolt. This stock Volkswagen item serves a dual purpose. It is used to hold the flywheel in place as well as acting as the pilot bearing for the transmission input shaft. The late model unit is recommended because it uses a needle type bearing instead of the bushing type that is found in the earlier Volkswagens.

These four simple steps complete the machine work necessary to make the engine conversion. Hopkins and Warren then reassembled the engine and installed Douglas dual mufflers made for the Corvair, first cutting away the crossover pipe and welding up the ends. This was necessary in order to clear the transmission.

Hopkins and Warren then designed an adaptor plate that would connect the Corvair engine to the Volkswagen transmission. Lukes & Shorman did the machine work on the adaptor casting and were so enthusiastic they decided to produce the adaptor plate in quantity for those interested in the conversion. It now can be purchased from Lukes & Shorman for $70. They will also do the necessary machine work to the Corvair crankshaft on an exchange basis for an additional $25.

The finished adaptor plate was added along with the VW flywheel to the Corvair engine. A VW Commercial clutch disc and a Porsche Carrera diaphragm pressure plate (Part #692) are used to transfer the torque.

Since the Volkswagen utilizes a 6-volt electrical system, whereas the Corvair is 12-volt, Hopkins and Warren felt it would be cheaper and simpler to convert the engine to 6-volt rather than to modify the VW's electrical system.

This entailed merely changing the generator and coil. The end mounting plates of the Corvair generator were retained. The case, armature and field were replaced with standard 6-volt Delco components. The coil is a 1954 Chevrolet 6-volt unit.

The VW transmission/differential unit was removed from the car and taken to the nearest Volkswagen dealer to have the ring gear flopped over to the other side of the case. Since the Corvair engine turns opposite to the Volkswagen, it is necessary to flop the ring gear, otherwise you would wind up with four speeds reverse and one speed forward. Fun!

The transmission/differential unit was then re-installed in the Ghia and the original carbon throw-out bearing was replaced with a later style unit.

The Volkswagen starter motor is attached to the transmission case and it was necessary to replace it, due to the

TOP — Putting a Corvair flat six into the space normally occupied by a small VW four isn't nearly as difficult as it might at first seem. All that Hopkins needed was an adaptor plate (casting at left) to replace the Corvair bell housing at right. A Porsche Carrera clutch handles the extra horses available from engine exchange. ABOVE — Engine about to be installed. Note rear rocker panel is removable, exhaust rerouted to clear stock Volks trans. This conversion has worked very well.

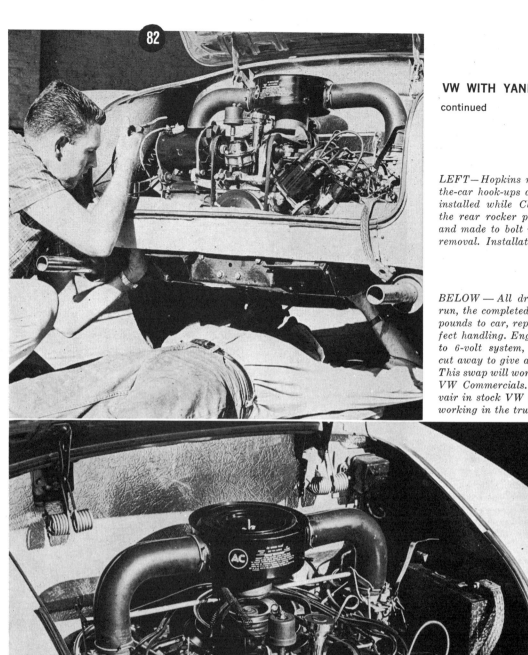

VW WITH YANKEE PUNCH
continued

*LEFT—Hopkins makes some under-
the-car hook-ups after the engine is
installed while Claude Warren fits
the rear rocker panel that was cut
and made to bolt in for easy engine
removal. Installation time is short.*

*BELOW — All dressed up ready to
run, the completed swap adds but 45
pounds to car, reportedly doesn't affect handling. Engine was converted
to 6-volt system, parts of flooring
cut away to give adequate clearance.
This swap will work OK in Ghias and
VW Commercials. To install a Corvair in stock VW means some metal
working in the trunk lid department.*

opposite rotation of the Corvair engine. It was found that the Bosch starter motor from the Goliath 4-cylinder car turns in the proper direction and with minor modifications can be used.

The armature was pulled out of the Goliath housing and compared with the VW unit. It was noted that the drive gear end of the armature that goes into the adaptor plate of the VW starter was smooth, whereas the Goliath armature shaft was threaded on this end. By chucking the Goliath armature in a lathe the threads were removed and the Goliath starter reassembled, substituting the VW adaptor plate which bolted right on. This unit was then installed in the original position on the transmission case.

Making a cardboard template of the outline of the Corvair engine, Hopkins and Warren placed it inside the engine compartment of the car and marked the area that had to be cut away. This was accomplished with an ordinary saber saw using metal cutting blades. To facilitate removal and reinstallation of the engine in the future, the rear rocker panel was cut and made to be removable.

The engine was then installed in the car. The original Volkswagen engine uses no motor mounts other than the bolts that hold it onto the transmission. With the Corvair engine weighing only an additional 45 pounds, it was decided to try doing without any additional motor mounts. In over 8000 miles no problems have arisen and the original mounts do the job satisfactorily.

In the Corvair, the gas hook up is on the same side as the Volkswagen, so the existing gas line in the car was retained and cut off approximately over number 6 cylinder, connecting to the Corvair fuel intake line by means of neoprene hose.

For the accelerator, the original Volkswagen unit is used. The late model Corvair throttle bar is used, with the actuating lever relocated 3 inches to the left of center. An additional lever is mounted ¼-inch away to form a connection to take the Volkswagen throttle set-screw unit. The throttle return spring runs from this lever to the generator case.

The original Volkswagen generator warning light was used by merely lengthening the wire and attaching it to the generator in the same manner as the original. Hopkins and Warren installed Stewart-Warner oil pressure and temperature gauges to the engine. They feel that this is a must and should not be left out.

This completed the installation and the car was now ready for road testing.

Author John O'Donnell took this photo just as the engine was being mated to the stock VW transmission/rear end. Ring gear was switched to the other side due to opposite rotation of Corvair engine. Adaptor plate is made by Lukes & Shorman Co.

To the smug satisfaction of the builders everything worked perfectly. This is indeed a rarity and goes to show the detailed thought that was put into the conversion.

On the quarter-mile strip the car has been clocked at a flat 80 mph with a respectable elapsed time of 17.35 seconds. Top speed is estimated at 115 mph. What would happen if a blower kit were installed? Or for that matter, any of the current hop up treatments incorporated to give the Corvair more punch? Better watch out for those Ghia's from now on, you might tangle with the wrong one!

Corvair throttle linkage was reworked slightly so the stock VW throttle connection could be used. No mounts other than stock ones on transmission are used to support engine. This conversion gives good performance. Think what a blower would do.

POOR MAN'S PORSCHE

Coming off the line at Irwindale (right) the EMPI *burned no rubber despite 105-hp engine due to suspension goodies and big rubber which give it too much bite. The hot Ghia was still much faster than standard, turned the quarter in the 17-second bracket. The Ghia looks very innocent (far right) with only the wheels and* EMPI *"bug" to create suspicion that all is not what it seems to be.*

The time was (not so long ago, really) when life held certain immutable truths; things in which a person could place his trust and faith. A Yankee victory in the World Series; Republicans always winning in Maine and Democrats always winning in the Solid South; Mae West's bosom; Volkswagens being reliable, but slow and ugly.

So the Yankees turned into a second division club, Republicans and Democrats win (or lose) elections in unaccustomed climes, while time, silicon, and a new generation of uninhibited starlets have turned Miss West into a memory. So all we had left to believe in was the ugliness and sluggishness of the VW. Messrs. Karmann and Ghia did in the former some years ago, and now EMPI (for Engineered Motor Products, Inc.) have put the kibosh on the latter.

With all these homilies being laid waste before our eyes, all we had left to hold on to was the laminated wooden steering wheel in the "Named by VW, Built by EMPI" Karmann-Ghia coupe that the Riverside, Calif., accessory and speed equipment firm placed at our disposal. It felt good. Then we fired up the 1677cc engine, spurred the 105 horses, listened to the sound in the bundle-of-snakes exhaust, and promptly lost interest in the Yankees, elections and Mae West. By the time we were out of the parking lot, we knew that this was going to be fun.

Our biggest objection to hopping up most small engines is that torque, economy and reliability all go out the window in the ruthless search for power. EMPI's chief mechanic, Dean Lowrey, was aware of this and very shrewdly prepared an engine that not only preserves these virtues, but improves them while more than doubling the 50 prancing Prussian ponies in the stock 1300cc engine (this was a 1966 car; the '67s are 1500cc, 53 hp). The extra cc's aid both low-range torque and horsepower. This is accomplished by using EMPI's own 88mm cylinder barrels and forged aluminum pistons. The camshaft, while more radical than stock, is the tamest of three offered by EMPI. Known as the ECSV-440 Sport Cam, it is meant to be a street grind. It produces very little power

below 2000 rpm, but from there on it really snaps the needle to the 5500 rpm red line. The EMPI-Speedwell Sprint Kit with two Stromberg 1½-inch variable-venturi carbs on special 2-piece manifolds and the EMPI Extractor exhaust system do a lot for inhaling and exhaling all through the range.

To be altogether fair, we got a 1967 standard 1500cc Ghia to drive and put it through the same paces as the EMPI. We knew it wouldn't be as peppy, but we got several other surprises as well. Compared to the old 1192 VWs, neither the new VW beetle (tested in the March '67 MT) nor the Ghia are slugs. In fact they both went and handled impressively. The EMPI version was not a better car; it was a wholly different car. How different? Well, diet cola versus sour mash seems a suitable comparison. We like soft drinks, but bourbon has the kick.

Power was less than half the story. Handling was another chapter, and EMPI's concern with little things that annoy even dedicated VW owners is the rest of the book.

Lowrey had mentioned when we took the car that it had their camber compensator on the rear end and a sway bar on the front, along with heavy-duty shocks. We had already noted the Speedwell-BRM wide-rim mag (no, not aluminum "mag," real 90% magnesium alloy) wheels and Pirelli Cinturato tires (165 x 15 fronts, 185 x 15 rears) and knew they would help a lot too. How much all this helped we only discovered when we drove through the mountains just after the snow had melted.

Does this package make the car handle? Did Errol Flynn like girls? The slight oversteer which we found present but not excessive or objectionable in the standard 1500 Ghia and Beetle all but disappeared in the EMPI. It would appear only if we hit a wet or sandy spot with the power on in a turn, but otherwise it was all gone. We were even able to induce a slight understeer by entering the turn well below optimum and then applying bags of power right at the entry to the bend and keeping it on all the way through. This technique also gave us optimum control at speed when entering strange

Ghia goodies galore! BRM-Speedwell wheels, made special for EMPI *in England, are rear mags, carry oversize Pirelli Cinturato tires. Top is a very realistic spray-on imitation vinyl. Adhesive strips simulate seam found in the real thing. Woodrim steering wheel, Slick-Shift, adhesive "wood" dash trim, and additional instruments are* EMPI's *cockpit additions. The sting in the tail is provided by special oversized barrels to increase displacement to 1677cc, dual Stromberg induction kit, full-flow oil filter, special oil cooler, special cam. Note hefty throttle linkage.*

turns, giving us some clue about what the bend looked like before we got in over our head.

Although that big rubber contributed mightily to the good handling of the EMPI-Ghia, it was a mixed blessing, as we discovered during the acceleration tests. The Pirellis had just a bit too much bite, bogging us down coming off the line. Just a wee bit of wheelspin would have permitted better times. Nonetheless, our test car got through the quarter 2.6 seconds faster than the stock version (17.9 versus 20.5) and its terminal velocity, 78 mph, was a 16 mph improvement over the tame one.

All this performance improvement should have had the needle on the gas gauge moving faster than Dillinger leaving a bank. Surprise! The EMPI not only wasn't a gas gobbler, but actually got better mileage than the stocker. Even in the mountains where we kept it wound up tight in 3rd most of the time and rarely used 4th seeking optimum performance, it got fully 1 mpg better than the showroom job which was driven more moderately. This economy edge was slightly improved on the expressway cruising at 65-70 mph. Part of this is due to the effect of longer gearing provided by the bigger rear tires, part by the more efficient engine.

EMPI has also paid attention to some of the other less obvious shortcomings of the VW and incorporated many of their corrective goodies in this car. For example, we've always been bugged by the long throws in the shift linkage which make us lean forward to get 3rd gear. EMPI cured this with their E-Z Shift conversion. Going from 1st to 2nd we thought it was hung up in neutral, before realizing that it was actually all the way home in 2nd. Lever movement is reduced 40% and it shifts quicker than a pit boss's eyes.

A true 100-mph-plus VW needs more than stock binders, so EMPI did something about that too. Their power brake kit with special linings is almost too effective; we found it just a bit sudden. It has a kind of "all-on all-off" sensation which made us rather careful about doing anything in a hurry that might make the car twitchy. In a straight line they stopped

things *toute suite,* though. Bad weather and a crowded schedule at the Irwindale Drag Strip, where we did the performance testing, foiled our plans to check actual stopping distance, but we're sure it would be better than standard, which is saying a lot.

A tach, oil temperature gauge and oil pressure gauge augment the usual instrumentation. The latter was quite handy as we discovered that oil pressure tended to come up quite slowly when the engine is fired up although the idiot light goes out instantly. The gauge probably saved the very unique engine one morning when we saw the presssure start up, then suddenly sag. We shut if off before the light came on and discovered that a fitting on the full-flow oil filter had broken and had dumped enough oil on our driveway to have groomed Rudoph Valentino's hair for 30 years.

All good things cost money, but this Super-Ghia is surprisingly moderate. The engine kit, for example, is about $600, not bad for a 100% increase. Virtually everything on the car could be duplicated for something in the neighborhood of $1500. This would make it a better performer than a Porsche 912 at less than the Porsche price. Another inducement is that this expenditure can be made in easy stages rather than in a lump sum. Also not to be overlooked is that Karmann-Ghias do not attract the gendarmerie's eyes as readily as Porsches do.

The EMPI-Ghia's most impressive feature, to us, was the restraint used in setting it up. Lowrey can get even more out of a VW engine, as his dragsters attest, but this car had none of the intractability of many high-performance cars. It is clearly a street machine, and we can vouch for its traffic manners. So can its lucky owner, the EMPI Office Manager. She drives it to work, the grocery store, the laundromat, etc., every day, except the days she let MOTOR TREND use it.

Of course, if you really want your Ghia to git, Lowrey is preparing a stroker kit to open it up to about 1900cc, plus other goodies that will raise the output to about 135 hp. Then he's going to surprise hell out of a Porsche 911.   /MT

PHOTOS/PAT BROLLIER

SCG ROAD TEST

Have you ever asked yourself...

A VOLKSWAGEN IS NOT AN AUTOMOBILE. A Volkswagen is the basic means of transportation for those people who don't need a car for its status, for its extension of their ego, or for the augmentation of their physical powers. As such, or as non-such, it really isn't logical to run a comparison test of it against an automobile. It would make as much sense to compare the telephone — a basic means of communication — to an automobile.

It is said that the happiest drivers in the world have Cadillacs and Volkswagens. That is, they are the owners who were rewarded far greater than their expectations. When you buy a Volkswagen, all you expect is an inexpensive and reliable vehicle for transporting yourself from point A to point B with a minimum of attention — either yours or your spectators'. Well, that's all it does, but it does that simple chore so straightforwardly that you begin to take it for granted that that is all a car is supposed to do.

The great majority of car owners, however, think differently. Since their car is their alter ego, it must also entertain, impress, coddle, carry five extra persons once or twice a year, run twice as fast as the legal limit, win every redlight-to-redlight drag race, haul more luggage than any ten people could rightfully need, on a two-week trip, and most important, be the ubiquitous, silent yardstick of their success. That's the majority. Presently. But the world's auto manufacturers are worried, and justifiably, because every year the standard of the "water-class" of cars nibbles off just a little larger share of the market. *Every year* — for a decade. Perhaps this says something about society that all the sensationalistic journalism and public opinion polls and demonstrations don't. I sure as hell hope so. With all due apology to certain theological groups, perhaps this is the beginning of the Forth Reich, the right Reich, a *social* Reich.

Dr. Ferdinand Porsche was without a doubt one of the ten most brilliant automotive engineers in history. With the Volkswagen and Porsche to his credit, few persons are even aware of his most advanced design, the Cisitalia, a contemporary with today's Formula 1 cars that was never developed after his death due to political complications. But now we can all wonder at whether it was managerial genius or sheer chance that retained and refined his original VW design for over two decades. The Beetle is outdated now, but the general public doesn't know that, just as they don't know that its proper successor, the F.I. sedan, has already arrived.

BEETLE

PRICE

Base ....$1935 (POE West Coast)
As tested $2280
Options ..Radio, whitewalls, air conditioning

ENGINE

Type ....Flat 4, air-cooled, aluminum block,
aluminum head
Displacement ....91.1 cu. in. (1493 cc)
Horsepower ....53 hp at 4200 rpm
Torque ....78 lbs.-ft. @ 2600 rpm
Bore & stroke ....3.27 in. x 2.72 in.
(83.0 mm x 69.0 mm)
Compression ratio ....7.5 to 1
Valve actuation ......Ohv, rocker actuated
Induction system ....Solex 30 PICT-2
Exhaust system ..Cast iron headers, 4 into 2
Electrical system....12-volt generator, point
distributor
Fuel ....Regular
Recommended redline ....4200 rpm

DRIVE TRAIN

Clutch ....Dry disc, cable actuated

Transmission	Gear Ratio	Overall Ratio
1st Synchro	.....3.80	.......15.68
2nd Synchro	.....2.06	.... 8.50
3rd Synchro	.....1.26	.... 5.20
4th Synchro	.....0.89	.... 3.67

Differential ....Spiral bevel, 4.125

CHASSIS

Frame .....Unit construction, rear engine,
rear drive
Front suspension ....Double trailing arms,
torsion bars, tube shocks, anti-roll bar
Rear suspension ..Single trailing arm, torsion bars, tube shocks
Steering ....Worm and gear, 2.8 turns,
turning circle 36.0 feet
Brakes ....Split hydraulic, all drum,
swept area 96.1 sq.in.
Wheels ....Steel disc, 15-in. dia.; 5-in. wide
Tires ....Continental 5.60 x 15 bias ply,
pressures F/R: 16/24 (rec.), 22/30 (test)

BODY

Type ....Unit steel, 2-door, 5-passenger
Seats ....Front buckets, rear bench
Windows ....2 manual, 2 vents
Luggage space ..Front & rear trunk, 8.6 cu.ft.
Instruments ....90 mph speedo
Gauges: fuel
Lights: gen., oil

WEIGHT AND MEASURES

Weight ....1900 lbs. (curb), 2130 lbs. (test)
Distribution F/R 39.5%/60.5%
Wheelbase ....94.5 in.
Track F/R ....51.6 in./53.3 in.
Height ....59.1 in.
Width ....61.0 in.
Length ....158.6 in.
Ground clearance ....5.9 in.
Oil capacity ....2.5 qt.
Fuel capacity ....10.6 gal.

MISCELLANEOUS

Weight/power ratio (curb/advertised) ...36.0
(test/dyno) 53.0
Advertised hp/cu.in. 0.58
Speed per 1000 rpm (top gear) ...19.8 mph
Warranty ....24 mos./24,000 miles

AERODYNAMIC FORCES AT 100 MPH
150 lbs. 0 lbs. 370 lbs.

CORNERING CONDITION
.70 g

CORNERING CONDITION
4.2°

ACCELERATION — g's
SPEED
ACCELERATION-g's
BRAKING-g's
SECONDS

PERFORMANCE

Acceleration ....0-30 (6.0 sec.) 0-60 (20.4 sec.)
0-quarter mile (21.5 sec., 63.0 mph)
Top speed ....78 mph (claimed) at 3940 rpm (factory limited)
Braking ....Distance from 60 mph: 164 ft. (0.73 g av.)
Number of stops to fade: Not attainable
Stability: Excellent
Maximum pitch angle: 0.5°
Handling ....Max. lateral: 0.68 g right, 0.70 g left
Skidpad understeer: 4.8° right, 6.6° left
Maximum roll angle: 4.2°
Reaction to throttle, full: more understeer; off: less understeer
Dynamometer ....Road horsepower: 40
Condition of tune: Carburetion slightly lean

Speedometer	30.0	40.0	50.0	60.0	70.0	80.0
Actual	28.0	37.5	45.5	55.5	65.5	77.0

Mileage ....Average: N.A.
Miles on car: 9000 to 9600
Aerodynamic forces at 100 mph:
Drag ....370 lbs. (includes tire drag)
Lift F/R ....150 lbs./0 lbs.

TEST EXPLANATIONS

Fade test is successive max. g stops from 60 mph each minute until wheels cannot be locked. Understeer is front minus rear tire slip angle at max. lateral on 200-ft. dia. Digitek skidpad. Autoscan chassis dynamometer supplied by Humble Oil.

SQUAREBACK

PRICE

Base ....$2629 (POE West Coast)
As tested $2723
Options ....Radio, whitewalls

ENGINE

Type ....Flat 4, air-cooled, aluminum block, aluminum head
Displacement ....96.7 cu.in. (1584 cc)
Horsepower ....65 hp at 4600 rpm
Torque ....87 lbs.-ft. @ 2800 rpm
Bore & stroke ....3.36 in. x 2.72 in. (85.0 mm x 69.0 mm)
Compression ratio ....7.7 to 1
Valve actuation ......Ohv, rocker actuated
Induction system ..Electronically timed fuel injection
Exhaust system ..Cast iron headers, 4 into 2
Electrical system ...12-volt generator, point distributor
Fuel ....Regular
Recommended redline ....4600 rpm

DRIVE TRAIN

Clutch ....Dry disc, cable actuated

Transmission	Gear Ratio	Overall Ratio
1st Synchro	3.80	15.68
2nd Synchro	2.06	8.50
3rd Synchro	1.26	5.20
4th Synchro	0.89	3.67

Differential ....Spiral bevel, 4.125

CHASSIS

Frame ......Unit construction, rear engine, rear drive
Front suspension ..Double trailing arms, torsion bars, tube shocks, anti-roll bar
Rear suspension ..Single trailing arm, torsion bars, tube shocks
Steering ....Worm and gear, 2.75 turns, turning circle 36.4 feet
Brakes ..Split hydraulic, disc front, drum rear, swept area 96.1 sq.in.
Wheels ....Steel disc, 15-in. dia.; 5-in. wide
Tires ....Dunlop 6.00 x 15 bias ply, pressures F/R: 17/26 (rec.), 22/31 (test)

BODY

Type ....Unit steel, 2-door, 5-passenger
Seats ....Front buckets, rear bench
Windows ....2 manual, 4 vents
Luggage space ..Front & rear trunk, 29 cu.ft.
Instruments ....100 mph speedo
Gauges: fuel
Lights: gen., oil

WEIGHT AND MEASURES

Weight ....2200 lbs. (curb), 2430 lbs. (test)
Distribution F/R 39.0%/61.0%
Wheelbase ....94.5 in.
Track F/R ....51.6 in./53.0 in.
Height ....57.9 in.
Width ....63.2 in.
Length ....166.3 in.
Ground clearance ....5.9 in.
Oil capacity ....2.5 qt.
Fuel capacity ....10.6 gal.

MISCELLANEOUS

Weight/power ratio (curb/advertised) ...34.0
(test/dyno) 56.0
Advertised hp/cu.in. 0.67
Speed per 1000 rpm (top gear) ....20.1 mph
Warranty ....24 mos./24,000 miles

AERODYNAMIC FORCES AT

CORNERING CONDITION

CORNERING CONDITION

PERFORMANCE

Acceleration .....0-30 (5.0 sec.) 0-60 (17.1 sec.)
0-quarter mile (20.5 sec., 65.2 mph)
Top speed ......84 mph (claimed) at 4180 rpm (factory limited)
Braking ......Distance from 60 mph: 156 ft. (0.77 g av.)
Number of stops to fade: Not attainable
Stability: Excellent
Maximum pitch angle: 0.5°
Handling ......Max. lateral: 0.65 g right, 0.68 g left
Skidpad understeer: 1.7° right, 5.2° left
Maximum roll angle: 3.1°
Reaction to throttle, full: oversteer; off: less understeer
Dynamometer ......Road horsepower: 43
Condition of tune: Excellent

Speedometer	30.0	40.0	50.0	60.0	70.0	80.0
Actual	27.5	38.0	49.5	59.5	70.0	80.0

Mileage ......Average: 22 mpg
Miles on car: 100 to 600
Aerodynamic forces at 100 mph:
Drag ......340 lbs. (includes tire drag)
Lift F/R ......60 lbs./70 lbs.

TEST EXPLANATIONS

Fade test is successive max. g stops from 60 mph each minute until wheels cannot be locked. Understeer is front minus rear tire slip angle at max. lateral on 200-ft. dia. Digitek skidpad. Autoscan chassis dynamometer supplied by Humble Oil.

KARMANN GHIA

PRICE

Base .... $2520 (POE West Coast)
As tested $2585
Options .... Radio

ENGINE

Type ....Flat 4, air-cooled, aluminum block, aluminum head
Displacement .... 91.1 cu.in. (1493 cc)
Horsepower .... 53 hp at 4200 rpm
Torque .... 78 lbs.-ft. @ 2600 rpm
Bore & stroke .... 3.27 in. x 2.72 in. (83.0 mm x 69.0 mm)
Compression ratio .... 7.5 to 1
Valve actuation ...... Ohv, rocker actuated
Induction system .... Solex 30 PICT-2
Exhaust system .. Cast iron headers, 4 into 2
Electrical system ...12-volt generator, point distributor
Fuel .... Regular
Recommended redline .... 4200 rpm

DRIVE TRAIN

Clutch .... Dry disc, cable actuated		
Transmission	Gear Ratio	Overall Ratio
1st Synchro	3.80	.... 15.68
2nd Synchro	2.06	.... 8.50
3rd Synchro	1.26	.... 5.20
4th Synchro	0.89	.... 3.67
Differential .... Spiral bevel, 4.125		

CHASSIS

Frame ...... Unit construction, rear engine, rear drive
Front suspension .. Double trailing arms, torsion bars, tube shocks, anti-roll bar
Rear suspension .. Single trailing arm, torsion bars, tube shocks
Steering .... Worm and gear, 2.8 turns, turning circle 36.0 feet
Brakes .. Split hydraulic, disc front, drum rear, swept area 96.1 sq. in.
Wheels .... Steel disc, 15-in. dia.; 5-in. wide
Tires .... Dunlop 5.60 x 15 bias ply, pressures F/R: 16/24 (rec.), 22/30 (test)

BODY

Type ....Unit steel, 2-door, 2 + 2 passenger
Seats .... Front buckets, rear bench
Windows .... 2 manual, 2 vents
Luggage space ...... front trunk, 3.7 cu.ft.
Instruments .... 90 mph speedo
Gauges: fuel
Lights: gen., oil

WEIGHT AND MEASURES

Weight ....1945 lbs. (curb), 2175 lbs. (test)
Distribution F/R 41.0%/59.0%
Wheelbase .... 94.5 in.
Track F/R .... 51.8 in./53.3 in.
Height .... 52.4 in.
Width .... 64.3 in.
Length .... 163.0 in.
Ground clearance .... 5.9 in.
Oil capacity .... 2.5 qt.
Fuel capacity .... 10.6 gal.

MISCELLANEOUS

Weight/power ratio (curb/advertised) ... 37.0
(test/dyno) 54.0
Advertised hp/cu. in. 0.58
Speed per 1000 rpm (top gear) ... 19.8 mph
Warranty .... 24 mos./24,000 miles

AERODYNAMIC FORCES AT 100 MPH — 160 lbs. — 70 lbs. — 320 lbs.

71 g — CORNERING CONDITION

4.1° — CORNERING CONDITION

ACCELERATION - g's

SPEED

ACCELERATION-g's

BRAKING-g's

SECONDS

PERFORMANCE

Acceleration .... 0-30 (5.4 sec.) 0-60 (19.1 sec.)
0-quarter mile (21.1 sec., 63.5 mph)
Top speed .... 82 mph (claimed) at 4150 rpm (factory limited)
Braking .... Distance from 60 mph: 152 ft. (0.79 g av.)
Number of stops to fade: Not attainable
Stability: Excellent
Maximum pitch angle: 0.5°
Handling .... Max. lateral: 0.70 g right, 0.71 g left
Skidpad understeer: 6.0° right, 7.2° left
Maximum roll angle: 4.1°
Reaction to throttle, full: more understeer; off: less understeer
Dynamometer .... Road horsepower: 40
Condition of tune: Excellent

Speedometer	30.0	40.0	50.0	60.0	70.0	80.0
Actual	28.5	37.5	46.5	56.5	66.0	76.0

Mileage .... Average: 24 mpg
Miles on car: 4400 to 5400
Aerodynamic forces at 100 mph:
Drag .... 320 lbs. (includes tire drag)
Lift F/R .... 160 lbs./70 lbs.

TEST EXPLANATIONS

Fade test is successive max. g stops from 60 mph each minute until wheels cannot be locked. Understeer is front minus rear tire slip angle at max. lateral on 200-ft. dia. Digitek skidpad. Autoscan chassis dynamometer supplied by Humble Oil.

WHY WOULD ANYONE BUY A VW?
Continued

But enough of this chit-chat. It you weren't a VW enthusiast you wouldn't have read this far, and you probably know that there's been nothing new in these cars for almost a year now. However, with our new electronic test equipment and road test procedure, we are now able to tell you some things about these cars that you've never heard before.

Tell Me Doctor, Is It Safe?

You've all seen newspaper photos of a little compact import, impact-welded to the front end of an unharmed semi-truck cab. Pure sheedy sensationalism. The people in the car were no deader than the truck driver would have been if he had tried to relocate a freeway overpass abutment. It's all a matter of scale — relatively speaking. A minor furor was created when the Feds released a movie of a collision they staged between a Volkswagen and a full-size Ford. The VW definitely got the worst of it. So what. It's an accepted scientific fact that when two objects of similar construction collide, damage is going to be inversely proportional to their relative mass. This produced the obvious response from an unknown satirical wit, "That proves it. We've got to get rid of them. We won't be safe until we've banned all full-size Fords from the highways." But walls smash trucks that hit full-size sedans that bump pony cars that collide with compacts that cream crampacts — that run over dogs.

Great fleas have little fleas,
Upon their back to bite 'em,
And little fleas have lesser fleas,
And so ad infinitum.
— Anonymous paraphrase of Jonathan Swift,
On Poetry, A Rhapsody **(1733)**

Does this mean that we all must have the same size vehicles running the same speed in the same direction with no obstacles? Let's look a little deeper. From a theoretical standpoint (simplification to absurdity) a human can survive a complete stop from freeway speed in 5 feet if properly "packaged." Volunteers have shown that 30-g decelerations can be tolerated for split seconds without permanent harm if the forces are distributed and there is no contact with solid objects. Therefore, if the impact can be controlled (the antithesis of "accident"), and the interior doesn't intrude, 5 feet of body crush-space would be adequate for a limp-away collision. In a VW sedan it is 5 feet from the front bumper to your knees. But don't try it! The engineers haven't quite reached the theoretical ultimate yet.

The other major approach to auto safety engineering is accident avoidance, and this is where Corvair was attacked. Chevrolet spent millions in proving that a rear-engined, swing-axle car — like the older VWs — was just as good as any other car in avoiding an uncontrollable condition, as long as you knew how to drive. But then both manufacturers recognized the

flaw in that reasoning, and revised the rear suspension to make it even harder to get bent out on a curve.

The problem with the original design was in its tendency toward "jacking." When a car with swing axles rises, the tire patches move inboard, making the tread narrower, or conversely, in a cornering condition the tire patches move inboard, causing the car to rise, causing more roll, etc. This is an unstable condition and eventually the car may simply rise up on tip-toes and fall over. They can't do that anymore. No *way*. The revised rear suspension is so foolproof that we had no fear of any maneuver on the skidpad, including broadslides, "J" turns and power oversteer. The Squareback was able to lift an inside front wheel due to its higher center of gravity, but was nowhere near going over. The only way you'll ever get one of these cars on its top is to drive it off the road and/or hit something while going sideways.

While on the skidpad, some other points redemonstrated themselves. In the first place, all three cornered faster to the left, with the driver's weight on the inside of the turn. Secondly, the higher the center of gravity, the slower the corner: Karmann, 33.0 mph (0.71 g); Beetle, 32.5 mph (0.70 g); Squareback, 32.0 mph (0.68 g), though there obviously wasn't much difference and *all* the Volkswagens were better than most of the cars we have tested, except the Lotus Elan +2, which did over 0.80 g.

Thirdly — listen — they all showed steady-state *understeer*. That's right, on asphalt at those speeds we were not able to make the Karmann or Beetle oversteer with either full or zero throttle, though the Squareback did with some coaxing. For some reason, probably front alignment, the 1600 Squareback had almost neutral steer in a right turn and could therefore be induced to "hang it out" with full throttle. So have no fear, your VW is legally a controllable vehicle. OK, you uncaped crusader, Nader, let's see you start something *now*. Just try.

They're pretty good at braking, too, with discs on the front of the Karmann and Squareback that haul them down at 0.79 g and 0.77 g for as long and as quick as you want to repeat 60 mph stops. The Beetle was a little worse at 0.73 g, but still no fade, and all were very controllable — by regulating pedal pressure in relation to tire squeal.

Nothing much exciting happened in the drag race, except that we got a lot of wheelspin in the Beetle. It's pretty easy if you don't know how — just take your eyes off a gas pumper while he takes off with your gas cap. Under acceleration the fuel runs out, down the fender, over the running board and under the right rear tire, and when you look back the sun has evaporated it, causing great ?-?-? for a while. This also caused one very exciting left turn oversteer before the problem was discovered and corrected.

Still, without smoking right-rear tires the race was as you might expect, the 1600-cc Squareback winning with a 20.5-second

e.t., the low-drag Karmann second with 21.1 seconds and the Beetle last at 21.5 seconds. The last two only had the 1500-cc engine, of course, but they weigh about 300 pounds less, too.

Our aerodynamic testing has really gone over big. Everyone wants to know how bad their car lifts and how its air drag compares with other similar cars. To reiterate, the drag figure we give is taken at about 70 mph — and includes rolling resistance (tires) — and is then extrapolated to 100 mph just for a convenient reference. To be more accurate we might try to subtract out mechanical drag, but *total* drag is still what matters on the road.

Surprise! The Karmann had the lowest drag figure at 320 pounds. But the Squareback was second with 340 pounds and the Beetle a bad last at 370 pounds. Maybe Mr. Kamm had something there after all. According to aerodynamic theory, the rear of the "bug" slopes away too rapidly and the airstream probably separates just below the roof line. Well, anyway, it *looks* more aerodynamic.

The Beetle also has a bad front/rear aero lift ratio. At freeway speeds, the front picks up about 75 pounds, and the rear, zero. In a curve, this means understeer, but in a crosswind, oh woe — you guessed it — instability. The Karmann is middle ground: same at the front, more lift in the rear, and finally, the Squareback is balanced with equal lift front and rear. The Squareback has more rear lift than the Beetle? Say, maybe those long-tailed Porsches are on to something good?

Top speed, as long as it's over 70, is our "care-less" test. The factory says Beetle, 78; Karmann, 82; and Squareback, 84 mph. Who are we to argue? After 30 seconds of acceleration we got to 70, 71 and 74 mph, respectively, if not respectfully. Oddly, the "factory" speeds correspond roughly to rpms at which occur both peak horsepower and the nominal redline point. And that speed is also called "cruising speed." So don't write in and tell us that yours goes faster — until you've had your speedo calibrated — and *then* don't tell us either. Until they have a class for VWs at Bonneville, our interest in those figures is exceedingly minimal. But however fast you drive, be careful of the speedo. In the Beetle and Karmann it was way off (on the safe side) and had a tendency to get delayed, reading differently going up from going down.

Our gas mileage, of course, wasn't what the factory predicted: 26.7 mpg for Beetle and Karmann, 26.1 for the F.I. Squareback, but you can imagine how Sports Car Graphic staffers drive in L.A., and the Beetle disqualified itself by running off at the filler spout. Still, we ranged from 22.2 to 26.8 in the Karmann, and the F.I. showed about 22 mpg.

Well, so much for the facts and figures. Since you're still with us you must be a masochist, so stick around and we'll give you our opinions now.

The Beetle is dead, but no one realizes that fact until they've had a chance to

WHY WOULD ANYONE BUY A VW?
Continued

drive the 1600 Squareback or Fastback. However, for $700 more, it seems that all you get is more space and 100 cc's. Driving the 1500s you get the feeling that the power is just "adequate," but in the 1600, the power is still just "adequate."

Shifting is a real nuisance, not only the amount required because of the low power and narrow range, but the clumsiness and washiness of the stick. Something deep down inside tells me their automatics are going to be very hot sellers. Another auto-trans promotional stunt is an over-weight clutch pedal that tempts you to shift into neutral at stop signs. All in all, you get the impression that you're doing something wrong when shifting, except that occasionally it goes right into the gear you intended it to.

Maverick made a big deal about having a smaller turning circle than VW, but it looks like that's not saying much. For a car with such a short wheelbase it's surprising to learn that even a Detroit pony car can turn more sharply, which brings up the pros and cons of big-diameter tires. The incongruity of 15-inch wheels on such a tiny car has become accepted on VW, but they obviously account for the limited front seat footroom and offset pedals caused by encroachment of the huge wheel wells. On similar-sized cars, the use of 14-, 13- and even 10-inch-diameter wheels permits more interior space and a tighter steering angle, but VW overweighs the advantages of more brake room, tire and fuel economy, and the off-road maneuverability and traction of larger diameter tires.

There weren't any complaints about the handling at freeway speeds in the Karmann and Beetle, but then we weren't able to arrange any crosswinds. On the other hand, the minimal understeer that was noticed in the Squareback on the skidpad resulted in a feeling of instability at speed. In other words, if you make a sudden maneuver, the car tries to overshoot as it rolls. Either more rear roll understeer or roll damping might be in order here. Still, no duress. As long as you drive sanely, you'll never get upset by its handling performance.

Sane driving is probably VW's biggest problem. We observed a terrible tendency to drive madly in these cars just because you could go flat-out most of the time without breaking any laws. You also get the impression that smallness can get you in anywhere — under semi-trailers, on sidewalks, in elevators. Something about these bugs brings out more "Mr. Hyde" than a street super-stock . . . probably the impression that you can get away with it, since VWs are as invisibly common as grass — the perfect spies' car. "More anonymous people drive Volkswagens than any other kind."

Returning to the inside, we find all manner of goods and bads. The first thing that grabs your eye in the Karmann is a genuine imitation wood-grain simulated paper-covered dash. That's the bad, the good is that it was so neatly folded inside the corners of the glovebox door that you hardly noticed. The one-hand seatbelt latch is the most convenient restraint of any car we've tested, and can be easily and safely buckled even while driving. And it's neat — it almost becomes a reflex action to flip it from the doorpost hanger to the anchor on the driveshaft tunnel. The bad: due to its anchor location, the diagonal shoulder harness is a real falling bra strap that tends to restrain only your left bicep. Visibility is average, good, very good as you proceed from Karmann to Squareback, but bad where the immovable rear view mirror can completely hide a car at 30 yards or a truck at 50 yards. If you are really concerned with what might be in that concealed zone, the easiest solution is to tilt the mirror to near vertical.

Ah, but what really concerns people is how much space you don't have on the inside — they say. Let's qualify that. The Karmann is roomy — it has as much interior space as' a Miura. The Squareback is cramped — it must have half the room of a Rolls Royce Silver Ghost. But the Beetle has much more interior volume than the Karmann, and in turn is half as big as the Squareback, which has the same seating arrangement plus luggage space. However, the Karmann is only a 2+2, while the Beetle is a 5+0, and the Squareback is a 5+. Therefore, how old is Charley's wife?

To be honestly objective, the front seat room in all three is quite adequate, but rear seat room in any is absurd. To get an average person in the jump seat of the Karmann means the driver has to slide forward until his knees join against the dash and his tie gets wound up in the steering wheel. The Squareback rear seaters at least have head room, though fat good that does their poor bent peds. But it does make a Great Experiment — after 22.3 miles, any 2 persons so subjected will have become either enemies or lovers.

Technical details of the cars aren't worth going into — there can't be anything duller than reading a description of the parts in a 20-year-old design (unless it's having to write it), but perhaps you'd be interested in one automotive engineer's opinion of that design. The passenger packaging is good for ordinary operation with an ordinary passenger load of two adults and occasional children, but there isn't enough sheet metal crush space for protection against unfair drivers who want to thrash you with a big 4-wheeled deadly weapon. And luggage space is sufficient for moderate trips, considering the miniaturization and dispos ability of clothing, plus it may be in the best location, considering front-end collision again, except that the SAE thinks the gas tank ought to be between the rear wheels for best protection. Suspension design was bloody bad until a few years ago when they modified the rear, so now it's only half bloody bad. The only redeeming feature up front is that it saves a little space, but the idea of leaning tires to the *outside* of a curve . . . well, suggest that to a bicycle rider. As for the driveline, *no hey mejores,* as our Southern neighbors say, meaning: "The horses are starving, officer." However, the location is perfect unless you think you are a battering ram, or unless you are a racer, in which case you might put the engine in *front* of the rear axle, and the rumor is out that VW/Porsche is building such a production car.

Air-cooled engines? Great for low horsepower and simplicity. Opposed four-cylinder layout? Great for space-saving. Electronic fuel injection? *Pow!* They're not the first to use a "computer" in a production car, but if they can get wide customer approval of such a system, friend, we're on our way to automatic *everythings.* Cry not, rugged individualists, you'll still have a go pedal, stop pedal and steering mechanism, but computer-controlled stability, emergency warning and evasion, performance optimizers, personal-comfort tailoring, etc., will give you added capabilities you cannot imagine. Technologically, they could all be incorporated today, except for cost, and cost depends on quantity, which depends on acceptance, and VW may be paving the way. Though apparently the real reason they did it was for better smog control, since there's no other discernible difference in performance from a carbureted engine.

Speaking of cost, don't let anyone tell you that any different feature on a VW is different because the parts are cheaper that way. By the time you amortize tooling and engineering over ten million identical units of *anything,* it doesn't matter how it was designed, the primary cost per unit can almost be based on pounds of raw material and labor in assembly. Cost of repair may be less, also, because of the lower parts inventory required, familiarity (to VW mechanics) of operations, and accessibility. The instrument panel assembly is brilliant in s   ic ty, just lay across the open front trunk snap out a piece of cardboard — and there the stuff is. I'd be willing to pay extra for such a feature on my Yankee machine.

O.K., let's wrap it up. These cars have drawbacks and deficiencies — a crank actually fell off one of our cars and a friend got one delivered with the rear alignment so bad that the tires were gone in 18,000 miles — but maybe it isn't fair to complain since you don't expect much in the first place. They don't have most of the faults we have found in other test cars — but then they don't have the virtues either. The purists may hate Bugs, and no one worse than a Porsche owner. But the fact remains, man, a Bug will get you across town just as fast, and probably cheaper, especially when you get pranged. Love 'em, hate 'em, or ignore 'em, but you've got to admire them, because the thing they are supposed to do — transport you — they do extraordinarily well.

VW Karmann-Ghia

1600 Karmann-Ghia Coupé

ENGINE CAPACITY 96.66 cu in, 1,584 cu cm
FUEL CONSUMPTION 31.7 m/imp gal, 26.4 m/US gal, 8.9 l × 100 km
SEATS 2 + 2   MAX SPEED 90.1 mph, 145 km/h
PRICE EX WORKS 9,145 marks

ENGINE rear, 4 stroke; cylinders: 4, horizontally opposed; bore and stroke: 3.37 × 2.72 in, 85.5 × 69 mm; engine capacity: 96.66 cu in, 1,584 cu cm; compression ratio: 7.7; max power (SAE): 65 hp at 4,600 rpm; max torque (SAE): 87 lb ft, 12 kg m at 2,800 rpm; max engine rpm: 5,000; specific power: 41 hp/l; cylinder block: cast iron liners with light alloy fins; cylinder head: light alloy; crankshaft bearings: 4; valves: 2 per cylinder, overhead, push-rods and rockers; camshafts: 1, central, lower; lubrication: gear pump, filter in sump, oil cooler; lubricating system capacity: 4.58 imp pt, 5.50 US pt, 2.6 l; carburation: 2 Solex 32 PDSIT downdraught carburettors; fuel feed: mechanical pump; cooling system: air-cooled.

TRANSMISSION driving wheels: rear; clutch: single dry plate; gearbox: mechanical; gears: 4 + reverse; synchromesh gears: I, II, III, IV; gearbox ratios: I 3.800, II 2.060, III 1.260, IV 0.890, rev 3.620; gear lever: central; final drive: spiral bevel; axle ratio: 4.125.

CHASSIS backbone platform, rear auxiliary frame; front suspension: independent, twin swinging longitudinal trailing arms, transverse torsion bars, anti-roll bar, telescopic dampers; rear suspension: independent, semi-trailing arms, transverse linkage by oblique swinging trailing arms, transverse torsion bars, telescopic dampers.

STEERING worm and roller, telescopic damper; turns of steering wheel lock to lock: 2.80.

BRAKES front disc (diameter 10.91 in, 277 mm), rear drum, dual circuit.

ELECTRICAL EQUIPMENT voltage: 12 V; battery: 36 Ah; generator type: dynamo, 200 W; ignition distributor: Bosch; headlamps: 4.

DIMENSIONS AND WEIGHT wheel base: 94.49 in, 2,400 mm; front track: 51.57 in, 1,310 mm; rear track: 52.99 in, 1,346 mm; overall length: 168.50 in, 4,280 mm; overall width: 63.78 in, 1,620 mm; overall height: 52.36 in, 1,330 mm; ground clearance: 5.90 in, 150 mm; dry weight: 2,073 lb, 940 kg; distribution of weight: 40.9% front axle, 59.1% rear axle; turning circle (between walls): 35.1 ft, 10.7 m; tyres: 6.00 × 15; fuel tank capacity: 8.8 imp gal, 10.6 US gal, 40 l.

BODY coupé; doors: 2; seats: 2 + 2; front seats: separate, adjustable backrests.

PERFORMANCE max speeds: 23 mph, 37 km/h in 1st gear; 42.3 mph, 68 km/h in 2nd gear; 69 mph, 111 km/h in 3rd gear; 90.1 mph, 145 km/h in 4th gear; power-weight ratio: 32 lb/hp, 14.5 kg/hp; carrying capacity: 860 lb, 390 kg; max gradient in 1st gear: 45%; acceleration: 0 — 50 mph (0 — 80 km/h) 11.5 sec; speed in top at 1,000 rpm: 19.6 mph, 31.5 km/h.

PRACTICAL INSTRUCTIONS fuel: 85 oct petrol; engine sump oil: 4.40 imp pt, 5.28 US pt, 2.5 l, SAE 20W-30, change every 3,100 miles, 5,000 km; gearbox and final drive oil: 5.28 imp pt, 6.34 US pt, 3 l, SAE 90, change every 31,100 miles, 50,000 km; greasing: every 6,200 miles, 10,000 km, 4 points; sparking plug type: 145°; tappet clearances: inlet 0.004 in, 0.10 mm, exhaust 0.004 in, 0.10 mm; valve timing: inlet opens 7°30' before tdc and closes 37° after bdc, exhaust opens 44°30' before bdc and closes 4° after tdc; normal tyre pressure: front 17 psi, 1.2 atm, rear 24 psi, 1.7 atm.

VARIATIONS AND OPTIONAL ACCESSORIES electronically controlled fuel injection system (constant pressure) with injectors in inlet pipes; electrically controlled sunshine roof; automatic gearbox, hydraulic torque convertor and planetary gears with 3 ratios (I 2.650, II 1.590, III 1, rev 1.800), possible manual selection, max ratio of convertor at stall 2.5, central selector lever, hypoid bevel final drive, 3.670 axle ratio, speed in direct drive at 1,000 rpm 19.6 mph, 31.5 km/h, max speeds (I) 36.7 mph, 59 km/h, (II) 61.5 mph, 99 km/h, (III) 87 mph, 140 km/h.

KARMANN GHIA

RT/TEST REPORT

Unchanged styling and three more horsepower join the fight against planned obsolescence.

Smooth lines of the Karmann Ghia date back to August, 1955, when it was first introduced in Germany. Ghia also tried a multi-lighted 1600 version but it was never successful.

The so-called "pretty Volkswagen" was designed by Ghia back in 1955 and the body is added to a stock Wolfsburg Beetle platform chassis by Karmann in Osnabrück so hence the name, Karmann-Ghia. Sales were slow at first, rising from 500 to 6,000 annually during the first four years. By 1965, though, they had climbed to 12,000 in the United States alone and are currently running at the rate of 27,500 annually. Of the two-seaters on the U.S. market, only Corvette does better, which is pretty fast company for a car that has been unchanged in appearance for 16 years.

Despite its timeless and sporty looks, the Ghia hardly rates as a sports car without extensive, unofficial modifications to its engine and suspension. It weighs 1,918 pounds, a figure 111 pounds heavier than the Beetle and the same as the new Super Beetle. It shares the same 60-horsepower engine as the other two and the current chassis is of the old type with torsion bar front suspension. Obviously superior aerodynamics, however, give it a true top speed of 86 mph, compared to 81 mph for the Beetles.

Aside from appearance, the main advantage of Ghia's styling is the car's relative freedom from wind buffeting. To get a little technical, the closer the aerodynamic center of pressure is to the center of gravity, the more stable will be the car. More or less as an accident of the Ghia shape, these two points are closer to each other than in any other Volkswagen model and this, plus the 7.1-inch lower height, makes it the most stable of the bunch. Put another way, its handling in strong crosswinds is acceptable.

For this reason alone, the Ghia should be the favored candidate for super-tuning of the engine and suspension and you'd think you'd see as

The Ghia is 7.1 inches lower than the Beetle but still retains the generous 5.9-inch ground clearance. The old-style front torsion bar suspension is used.

many of them if not more on the drag strip than Beetles. Such, however, is not the case, perhaps due to first cost which is $2,420 at East or Gulf Coast ports of entry compared to $1,780 for the old-style Beetle. The engine is far more accessible than in any other Volks, there is more room for bolt-on hardware and the battery for once is the first thing to come to hand upon opening the compartment. You don't need a professional laryngoscope to check the electrolyte, and a special pitcher to top it off if necessary.

Depreciation-wise, a Ghia drops $238 the minute you drive it out of the showroom whereas a Beetle technically gains $31 in value. At the end of the first year, you're out $442 with the Ghia and only $67 with the Beetle. However, Beetles have traditionally commanded an unrealistic premium on the used-car market and the figures for the Ghia, when compared to other imports or U.S. domestics of the same initial cost, are not bad at all.

Our test Ghia was loaned to us by Economotors VW of Riverside, Calif., which is operated by the same people who run EMPI, the well-known speed equipment and accessory house. A strictly stock VW of any type emerging from Economotors into the hands of a customer is something of a rarity and for the same reason, the firm sells almost as many new Volkswagens to people from out-of-state as they do to Riverside area residents. It's a fine place to blow your mind and wallet on the extensive EMPI catalog of

goodies. Ours, for example, was equipped with special wheels, radial tires, a sports steering wheel and cocoa floormat for starters which, including the air conditioning, raised the sticker price by nearly $1,000. It turned heads despite its 16-year-old shape.

They did not, though, find time to do anything to the engine which in ultimate EMPI-ized form, as installed in the "Inch Pincher" Beetle dragster, has been known to shave 105 mph and 13 seconds in the quarter. Also, the Inch Pincher is not air-conditioned in the normal sense so our encumbered Ghia was not taken out to embarrass itself at the strip. On a quiet, straight country road it would do a legal zero to 60 mph with two up in 18.5 seconds and a slightly illegal 0-75 in just under 30 seconds. The three horsepower increase for 1970 makes this the first Ghia to get to cruising speed in less than 20 seconds and, of course, it would have done somewhat better without the air conditioning and with a little break-in. Despite Volkswagen's assurance that you can drive flat-out through the showroom door without harming the engine, these cars need a few thousand miles on the odometer to find their full potential.

Thanks to the lower height, cornering is quite noticeably better than the plain Beetle and about on a par with the MacPherson-sprung Super Beetle. We're talking here of normal maneuvering without discomfitting passengers, not running a road course. To our knowledge, Ghias aren't normally entered in any event more serious than a gymkhana and we evaluated it accordingly, the way a purchaser would normally use his car.

We left the air conditioner off

The Ghia's interior detailing was the first so far to rate nearly perfect in our ratings. Bodies are hand-assembled by Volkswagen's Karmann subsidiary.

Luggage space in front is a bit spartan but the spare is easy to get at. It also provides pressure to operate the windshield squirters as on the Beetle.

EMPI's styled wheels are particularly pleasing on the Ghia and weigh considerably less than the stock perforated units complete with separate cap.

Fairly brisk cornering produces little lean and the rear axle arrangement new in 1970 on stick shift models militates against violent oversteer.

isn't much room behind them and the edge of the luggage shelf. There are no pretensions made that the space in back is for humans, however small, as it is just carpeted, not cushioned. The body itself is hand welded, hand leaded and then hand sanded prior to being hand painted, which avoids the touch of pebble-grain served up by Volkswagen's automated electrostatic booths at Wolfsburg.

A Ghia bonus is the standard front disc brakes. As we had borrowed the car from a dealer who already had a customer for it, we had no intention of damaging the expensive radials in a panic stop. Thus, our figure of 146-feet from 60 mph could undoubtedly be improved upon but it is still respectable. We also only tried the exercise once instead of the usual five successive times for the same reasons of respect for another person's property. We presume fade

throughout the test so that it would not adversely affect fuel economy any more than necessary. Even so, the extra weight if nothing else cut us back to 20.4 mpg in the city. On freeways and in the country the sleeker shape came into its own, giving us 26.3 mpg which is about one mpg less thirst than a Super Beetle's and quite acceptable by economy import standards. Oddly, people think nothing of paying a whopping surcharge for a Ghia or Type III but they insist on the usual Volkswagen economy.

Karmann does the interiors as well as the body, and the finish of both rates the highest marks we have yet to give a car of any price. Perhaps practice does count but it's still satisfying to sit in a new car and be unable to find a flaw—not only satisfying but extremely rare. The seats give you better back support than stock Wolfsburg buckets and the headrests are of a reasonable size. The backs will recline to a limited extent but there

Ghia front ends are a bit vulnerable in parking lots despite standard bumper guards so a business has grown around auxilliary center grilles to cover the damage.

would be no problem with the discs but aren't sure. It was, for example, a bit in evidence with the similar system used on our test 411.

Instrumentation is limited to speedometer, clock and fuel gauge per standard Wolfsburg practice but there's plenty of room on the walnut veneer dash to put aftermarket engine gauges in a visible and logical position without sacrificing your radio speaker as on the Beetle. The controls for the neatly hung-on air conditioner are super simple, consisting of one switch for the fan and one for temperature. This is not integrated with the old-style, non-flow-through ventilating and heating system still used for the Ghia. (Who'd want to cut vents in the rear quarters of this lovely body?).

Steering is a quick 2.7 turns, lock-to-lock, and the U-turn capability is still the same reasonable 36 feet as the torsion-bar Beetle. The Ghia probably wasn't given the MacPherson coil

The Ghia's 60-horsepower, 1,600cc engine is the most accessible of all Volkswagens. Note the practical location for the battery where it can get regular attention.

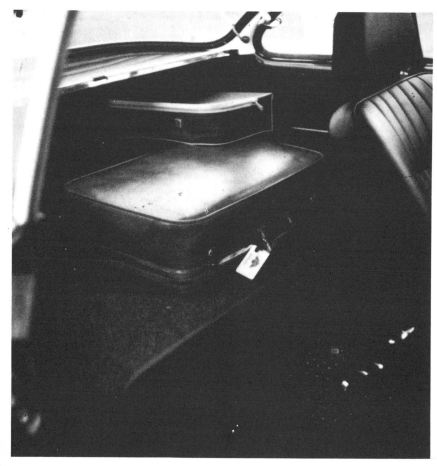

There's more room for luggage behind the seats. This compartment is carpeted but not padded as it's not intended that even small children ride there.

find much good to say about this Ferodo device and neither, apparently, can the public because not many are sold, relatively speaking. Progress, when the clutchless stick is left in high, is painfully slow and not a recommended technique in traffic anyway. Then when you do shift, just a slight pressure on the stick from your hand after the new gear is engaged will leave you and the engine churning futilely in the middle of an intersection. This can also happen if you or the passenger inadvertently hit the thing with a knee. We tested one once, months ago, and never did get used to it in 1,000 miles although undoubtedly owners finally do.

But each to his own. We'd like to buy a Karmann Ghia from Economotors, drive it around the block to make a used car of it, and then turn the EMPI people loose on it. For about $400 in parts and labor, you can convert any Volks except maybe a bus into a machine that will do zero to 60 mph in 12 seconds or so and still be as durable and as tractible in traffic as a Wolfsburg original. Then, all you need would be another $40 for an EMPI or Hurst shifter and you'd have a poor man's air-cooled Ferrari.

Don MacDonald

system because there is no particular pressure to increase luggage capacity or correct wind bobble in this strictly two-passenger car.

The convertible at $2,630 is continued in the line for '71 with its typically Teutonic top that folds neatly like a mattress behind and above the luggage compartment. We don't know another example of this fast diminishing breed where the sur-charge over a steel roof is less, not even the Beetle's version which, incidentally, is not available in Super Beetle form. Unfortunately, however, both types depreciate at a faster rate than their closed counterparts and it's expensive to duplicate the factory tops when replacement is necessary.

The major Ghia option, aside from air conditioning, is the semi-automatic transmission at $139 extra. We can't

Discreet badging – the name of the car comes from those of two coachbuilders, Karmann of Osnabrück and Ghia of Turin

Wilhelm Karmann – company boss and son of its founder

Collectable Karmanns

The Karmann-Ghia was a case of turning an ugly Beetle into a beautiful butterfly. Anders Clausager chronicles its history and points out its virtues and vices

Arguably, the Volkswagen Karmann-Ghia is neither classic nor sports; but it is equally far from being a red herring. Age sometimes gives respectability – the Wolseley Hornet Special and the Ford Consul Capri are both cases in point, and so is the Karmann-Ghia. All three might be described as boulevard sports cars, which, as a concept, is time-honoured.

The specification of the Karmann-Ghia would make any present-day motor industry product planner green with envy. Take one bog-standard motor car, do absolutely nothing to the mechanical specification, dress it up in a stunning new body (with less room than your standard model), and sell it at anything up to 60 per cent more.

Previous attempts at a pretty Beetle

Clothes make a man but can also make a motor car. Volkswagen's dear old Plain Jane Beetle was surely nobody's favourite styling exercise, and in its heyday always sold despite, rather than because of, its looks. As VW advertising was so fond of telling us, a pretty face is not everything. All the more surprising, therefore, that Volkswagen should give Jane a party frock. There had been several attempts at this, going back to the forties, such as the rare Hebmüller two-seater drophead, and the West Berlin-built Rometsch popularly known as the "Banana". But among the would-be Beetle customisers, Karmann coach-builder of Osnabrück enjoyed a special relationship with Volkswagen.

Wilhelm Karmann built his first car body in 1902 and by the twenties was thinking in terms of thousands of bodies rather than one-offs. Most of the company's output in the thirties was for Adler, and Karmann invested in tooling to mass-produce pressed-steel bodywork. When Adler did not re-start car production after the war, Karmann looked for business elsewhere.

At the time Volkswagen already had the Hebmüller-bodied two-seater convertible which was a factory-approved special, but many customers preferred a four-seater, and Karmann developed a four-seater convertible body for the Beetle chassis using a large proportion of standard panels. This was offered from 1949, alongside the Hebmüller model, though the latter was soon dropped leaving the Karmann Cabriolet ruling the roost. This it did most successfully. The Karmann Cabriolet remained in production for 30 years and was the last version of the Beetle to be manufactured in Germany. Chassis were supplied from Wolfsburg to Osnabrück where bodies were built and mounted – an arrangement which left the Wolfsburg factory free to concentrate on the saloon model. The Karmann Cabriolet was sold through the VW dealer network internationally and carried the company's usual warranty. It also sold at a premium of one-third more than the saloon model.

However, it still *looked* like the Beetle, so in due course Karmann found themselves considering the idea of building a car radically different in looks to the VW. A collaboration with the Italian design consultancy Ghia produced the answer – a svelte if voluptuous coupé body, strictly a two-plus-two. In fact the rear seats were only suitable for very small children and were more useful in their alternative role as additional luggage space. The really sensational feature was its very low build, cutting 7ins off the Beetle's 5ft height. It was also a few inches longer and wider . . . and almost 2cwt heavier.

Better aerodynamics

Mechanically the Karmann-Ghia was absolutely stock, except for a front anti-roll bar which only became standard equipment on the Beetle saloon in 1959. The engine was still the classic 1192cc size and developed all of 30bhp; but the weight increase was happily off-set by a reduced frontal area and better aerodynamics, so the Karmann-Ghia had a top speed of 74mph – 4mph more than the Beetle. Acceleration from 0 to 62mph was cut from 38secs to a still hardly neck-snapping 33secs, while fuel consumption was unchanged. The cost in Germany was DM7500 when the model was introduced in 1955. At the time, the de-luxe "Export" saloon cost DM4600 and the Beetle Cabriolet DM5990. For the parsimonious Volkswagen offered the standard saloon (almost invariably a home market special) at DM3790. In later years, however, the price gap between the Beetle and the Karmann-Ghia was narrowed.

Inevitably, the Karmann-Ghia was greeted with derision in many quarters. Obviously, it lacked the functional qualities of the Beetle, and equally obviously it was not a sports car. The Germans themselves rather unkindly referred to it as *Sekretärin-Auto* – a secretary's car. The Swiss

Volkswagen's own attempt

The VW Type 3 (right) was a straight development of the saloon. The car was completed in July 1961, but it was sent to the Frankfurt Motor Show for its first public appearance which interrupted development testing. After the show testing continued but, alas, various weaknesses appeared in the body structure. Righting the wrongs would have required a large investment, so the project was dropped.

How many convertible Type 3s were made? The answer is probably two. One was painted red and the other white. The red one is now on show at the Factory Museum, the white one used to be owned by the Frankfurt (Main) VW dealer Autohaus Glocker, but unfortunately after the original 1500 N engine was replaced by a Porsche unit, the car was involved in an accident and written off.

The Type 3, with its razor-edge swage lines and quad headlamps, had lost some of the stylistic flair of the Type 1

Out and about in a Type 1 convertible. When introduced to the UK the price was around that of the Austin Healey 3000

post-1959 Karmann-Ghia, with the raised headlights and larger 'nostril' air intakes

Automobil-Revue said, rather carefully, that 'in the opinion of many people, it is the best-looking car of its size and price'. Apparently there were more than enough people who agreed, because Volkswagen never had much trouble selling the Karmann-Ghia; they realised that most sales were probably made to confirmed Volkswagen enthusiasts who knew just what they were getting.

A beautiful but expensive car

About 40 per cent of the output went to the USA although the home market was always the most important outlet. British Beetle fanatics had to wait until the 1956 Motor Show before they could feast their eyes on the Karmann-Ghia. Prominent among them was Bill Boddy of *Motor Sport*, who was in raptures over 'one of the most beautiful cars at Earls Court . . . the striking new Karmann-Ghia'. He appeared unperturbed that the price was quoted at £1216 (inclusive of importy duty and purchase tax) when the Beetle saloon de-luxe cost £740 and the Beetle Cabriolet £1006. Leather upholstery was apparently a feature but right-hand drive was not always available in the early years of the Karmann-Ghia's career in the UK.

Much of the success of the Karmann-Ghia must be attributed to the fact that it was virtually first in the field of boulevard sports cars after the war. It offered a sensible alternative to dowdy-looking saloon cars, but was sensible in the way that it did not have the disadvantages of real sports cars. It was closed; it had room for the children, or for luggage; it did not cost the earth to insure, run and maintain; and spares were available just around the corner, anywhere in the world. But it did offer elegance, style, and a well-appointed and well-finished interior. The recipe would soon be copied by Renault with their Floride model, and then by many other car makers with varying degrees of success. Since 1969 the market leader has been Ford's Capri, but Volkswagen's own Scirocco is, in many ways, aimed at the same type of customer who would have considered a Karmann-Ghia 25 years ago.

The original Karmann-Ghia coupé was supplemented by a drophead version in 1957, and in 1959 both models were given a minor facelift; headlamps were raised, and the 'nostril' air intakes in the front panel were increased in size. The early models made before 1959 are now particularly prized among collectors for their bodies were hand-finished to a much higher degree than was the case for the later cars, and production was limited. The drophead coupé was first launched in Britain in 1958, at £1395. In the days before the Common Market, the import duty on foreign cars was pretty hefty and the Karmann-Ghias found themselves pitched on price against such British cars as the Austin-Healey 3000, the Daimler SP250 and the MGA Twin-Cam, to name but three home-grown sports cars; more direct competition might come from the Sunbeam Rapier, and later the Ford Consul Capri, and both were £200 less.

Small improvements made annually

Developments over the years followed the evolution of the Beetle. In 1960 the 34bhp engine was fitted, with a fully-synchronized gearbox, and five years later the 1.3-litre 40bhp engine was standardized. That only lasted a year as in 1966 the 1.5-litre engine from the "Super-Beetle" found its way into the back of the Karmann-Ghia; other improvements were dual circuit brakes with discs at the front. In 1967, an optional semi-automatic gearbox was offered in conjunction with the double-jointed swing axles which effected some improvement on the typical Beetle behaviour.

Small improvements were made annually; in 1970, the engine was again uprated, now to a full 50bhp 1.6-litre unit as used in the 1302S; the final edition from 1971 had impact-resistant bumpers and bigger rear lamps, neither of which improved the looks. Strangely, the manual gearbox version had simple swing axles to the end. Production ran out in 1973, after a total of 364,401 coupés and 80,899 convertibles had been made. Within a few months, the Karmann-Ghia had been replaced by the

Possibly the most desirable of the Karmann-Ghias, a Type 1 Convertible. Early ones are more collectable but this late, seventies, example offers rather more performance

Beneath the 'boot lid' there is the familiar VW flat four, all but hidden by its cooling ducting

A modern steering wheel, controls and trim look incongruous against a mid-fifties facia

Scirocco on the Karmann assembly lines as well as in the VW showrooms.

Meanwhile, Volkswagen had tried their best to make history repeat itself. The Type 3 1500 was introduced in 1961; its pancake engine permitted a rear luggage boot above the engine, and an estate car version was possible. The body styling was up-to-date but completely anonymous compared to the Beetle. The biggest mistake was that the Type 3 shared the Beetle's wheelbase of 2400mm (about 8ft) which pleased the production planners at Wolfsburg (who to this day have insisted on the same dimension for the wheelbase of the Golf and Scirocco!) but it meant that the 1500 offered about the same interior space as the Beetle, and a four-door model was not practical. The Type 3 was never a success by Volkswagen's standards; of 2.5 million cars made, half were estates ('Variants'), and most of the rest were split between the notchback saloon and the later fastback version. A convertible model had been shown in 1961 but was chiefly remarkable for its lack of torsional stiffness, so was hurriedly withdrawn; two prototypes were built, one of which survives to this day in the VW Museum.

By contrast, a Type 3 Karmann-Ghia coupé *did* get off the ground, and exactly 42,563 were built from 1962 to 1969. The recipe was as before – a Ghia design built by Karmann, but this time sadly lacking in stylistic flair. Where the original Karmann-Ghia was curvaceous, the Type 3 was all awkward angles and sharp edges, and sported a most peculiar-looking four headlamp layout. With the original 45bhp engine, top speed was a modest 85mph, so after a year a two-carburettor engine developing another 9bhp was installed, which gave a top speed of 94mph. In 1965 engine size was increased from 1493cc to 1584cc but output and top speed were not improved. In 1967 a fully-automatic gearbox became optional; as on the Type 1, the automatic model had double-jointed swing axles, but on the Type 3 these were standardized on the manual model in 1968.

That year also saw the introduction of another option, an electronic fuel injection system which remarkably appears not to have increased performance, nor decreased fuel consumption. The Type 3 Karmann-Ghia was discontinued in 1969, but saloon and estate versions of the VW 1600 remained in production until 1973. There was no official drophead coupé version of the Type 3 Karmann-Ghia but a few cars were individually converted. The model was listed in the UK from the 1963 Motor Show, originally priced at £1330; six years later this had risen to £1542. By comparison, the Type 1 Karmann-Ghia Coupé had gone up in price to £2263 by the time it was discontinued in 1973.

Collectors go for early models

Thanks to the numbers made, the Type 1 Karmann-Ghia coupé is not – and never will be – a rare car, although it is unusual in the UK as in most other countries outside Germany and the USA. From the collector's point of view, the 1955-1959 models are the ones to go for; but if you want some pretensions to performance to match the looks, go for the late 1.6-litre models which will almost, if not quite, see 90mph. They all share the virtues – and the shortcomings – of all other Beetles.

On the plus side, list quality and reliability; on the minus side, the handling and that bloody noise. The Karmann bodywork is said to suffer more from rust than VW's own. Much the same goes for the drophead version, with the added attraction of wind in the hair, and scarcity value. Mechanical spares should not represent any serious problem on either model; but body and trim parts are quite another matter.

The Type 3 Karmann-Ghia is a different kettle of fish altogether; it is what some classified advertisements would euphemistically describe as "a real collector's item"; a damning but faint praise which normally means that the car is not particularly desirable but its rarity is being used as an excuse to up the asking price. This model is for the devoted VW enthusiast who appreciates something, well, different if you see what I mean . . . but seek and ye shall find. Amazingly, there are still a few around in the UK. Parts are naturally proportionately more difficult than for the Type 1. And if the £1400 to £1700 bracket represents the approximate value of a good Type 1 Karmann-Ghia coupé, expect to pay rather more for the drophead but rather less for the Type 3 in similar condition. As usual, cars for restoration are at small fractions of these guide prices.

For those who prefer to stick to mantelpiece motoring, there is an excellent Dinky Toys model of the Type 1 Karmann-Ghia available in both British and French-made versions (turn it downside up to check – the French model is rarer in the UK). Another die-cast is the much rarer German Märklin which is to a slightly smaller scale, around 1/48th; and Revell made a plastic kit of the Type 1 in their Cadet series. The Type 3 Karmann-Ghia was modelled by Corgi Toys and this model features sundry opening lids.

The Karmann-Ghia name survived for several years on a Brazilian-built Volkswagen, the Karmann Ghia 1600TC which was made until 1975; apart from the name this had nothing to do with the German models, as it had a completely different body style which seemed equally remote from Italian design flair and German coachbuilding quality. The Ghia company is now the property of one of Volkswagen's arch competitors in the world markets, which effectively prevents VW from using Ghia's services and name. But the association with Karmann continues to flourish; the Osnabrück factory build the Scirocco, and also the Golf Cabriolet bodies for Volkswagen. Karmann, which celebrated its centenary in 1974, is not tied to Volkswagen however, and in recent years has also worked for Opel, Porsche and BMW. Indeed, they currently build the 6-series coupé body for BMW.